Biotechnology Unzipped

Promises and Realities

BIOTECHNOLOGY UNZIPPED

PROMISES & REALITIES

ERIC S. GRACE

Trifolium Books Inc.

Toronto

To Guy

Trifolium Books Inc.
250 Merton Street, Suite 203
Toronto, Ontario, Canada M4S 1B1

Canadian Cataloguing in Publication Data
Grace, Eric S., 1948-
Biotechnology unzipped: promises & realities

Includes bibliographical references and index.
ISBN 1-895579-45-7

1. Biotechnology I. Title.

TP248.215.G72 1997 660'.6 C96-932544-4

Trifolium Books Inc. acknowledges with gratitude the generous support of the Government of Canada's Book Publishing Industry Development Program (BPIDP). Support of the Canadian Institute of Biotechnology in the production of this book was also appreciated.

Printed and bound in Canada
10 9 8 7 6 5 4 3

Editor: Susan Lawrence
Project coordination: Diane Klim
Production coordination: Francine Geraci
Cover and text design: Blair Kerrigan/Glyphics
Illustration: Heather Collins/Glyphics

Ordering Information
Orders by Canadian trade bookstores, wholesalers, individuals, organizations, and educational institutions: Please contact General Distribution Services, 325 Humber College Blvd, Etobicoke, ON, M9W 7C3; tel. Ontario and Quebec (800) 387-0141, all other provinces (800) 387-0172; fax (416) 213-1917.

Trifolium's book may also be purchased in bulk for educational, business, or promotional use. For information, please telephone (416) 483-7211 or write: Special Sales, Trifolium Books Inc., 250 Merton Street, Suite 203, Toronto, Ontario, M4S 1B1.

Contents

Chapter 6

Chapter 7

Preface

Over the past year or so, when I've told people I'm working on a book about biotechnology, the usual response has been a cautious silence. Then: "What exactly is biotechnology?" they ask. "You know," I encourage them, "genetic engineering and all that." At this point, it's my turn for a thoughtful pause.

Biotechnology is a difficult subject to get one's head around. It's big, technical, and ever-changing. It isn't easy to explain what biotechnology is, and what it is not, in only a few sentences. The subject catches the imagination, but soon withers as a topic of conversation for lack of quick-to-communicate, easily understood facts. Then, too, biotechnology involves manipulations of living things that can seem bizarre, almost unbelievable, to the uninitiated. To tell of scientists transplanting genes from one species to another still sounds uncomfortably like either a horror story or the start of a bad joke. ("What do you get when you cross a hyena with a parrot?")

As a science writer, I followed news of developments in the field over the years in a back-of-the-mind sort of way. When a phone call from the publisher invited me to write about biotechnology, I thought it would be a good chance to polish up my learning. Had I realized just how much material has been published on the subject, I might not have been so keen. The flood of new research is almost overwhelming. Shelf after shelf of books and journals in college and university libraries deal with topics that I had, somehow, to squeeze into single chapters for this book, and the initial challenge was to decide how to organize and condense so much information.

This book, then, is the story of my discoveries. My aim at the outset was to get answers to some simple questions about a complex and wide-ranging enterprise. Looking at what was already published, I found some good introductory books about biotechnology written during the early years of its development, in the 1970s and 1980s. They describe the scientific basis of biotechnology (which hasn't changed) and the techniques it uses (which have had some important additions). The most obvious deficiency of those books for today's readers is their coverage of applications. Ten, or even five years ago, writers had to speculate about the many directions this new technology might take. Today, there's less question of what might be done. The foundations of the biotech business have now been dug, and the walls of commercial products and processes are rising staggeringly quickly. Although the details continue to expand and fill in, the outlines of the first success stories are now clear.

Given the rapid growth of biotechnology, and the public concerns about it, I resolved that this book needs to address three key questions: What is biotechnology? How is it being used? And what are some of the issues it raises? As you read through these pages, you will gain some pretty fair answers to all three questions.

The first chapter outlines the scientific background that made biotechnology possible — a primer of basic biology for understanding why biotechnologists can do the sorts of things they do. The second chapter explains some of the processes and procedures of biotechnology — the "how" of genetic manipulation. The bulk of the book, Chapters 3 to 6, explores applications of biotechnology, with examples of what is being done in the fields of medicine and agriculture (by far the two biggest areas where biotechnology is now being used), and in pollution control, mining, energy production, aquaculture, and forestry. The final chapter covers key issues: public attitudes to biotechnology, the patenting of genetically modified organisms, and the role of science and technology in society.

I approached the sometimes controversial issues in biotechnology without strong views either for or against. My personal fascination at first was less with the applications of biotechnology than with the *science* of it, and the startling evidence it gives us about the fundamental unity of all living things. My goal was simply to find out what was going on, using as many different sources as possible, and to make my own interpretations as I went along. I wanted to share my new-found knowledge with others who may not have had the time, inclination, or science background needed to plow through the often-complex literature on the subject.

Like any member of the general public, I'd seen newspaper headlines such as "Canola genes altered for profit" or "Genetic experiment cuts cancer spread." Soon, I was poring over science journals and reports from pharmaceutical and agribusiness companies, and being lured along the labyrinthine pathways of the Internet to hundreds of government statements, industry profiles, and patent descriptions. I read newsletters and books from groups opposed to some of the uses of biotechnology. I went to meetings — both pro and con — and spoke to scientists, businesspeople, and opponents of biotechnology applications.

It was a reasonably straightforward matter to find information about the techniques of biotechnology — such as how to cut and splice DNA, or clone genetic material, or make transgenic organisms (animals, plants, or microbes carrying the active genes of other species). Nor was there any shortage of details about where and how these techniques are being applied. The difficult part was evaluating the final outcomes, separating facts from speculation, and science from politics. Is biotechnology really bringing us cheap and effective drugs and disease-free crops, as many biotech companies promise? Or is it posing threats to the environment, human health, and animal welfare, as some citizens' groups contend?

It's typical of new technologies affecting society at large to attract both allies and enemies. The battle lines are drawn. Those who champion novel inventions as signs of a better future are on one side; those who warn about the dangers of sailing in uncharted waters are on the other. I give examples of both points of view throughout the book, with the arguments used by each side.

Our views about biotechnology, especially when it involves gene transfer, are closely tied to our perception of its goals. Seen as a way of increasing food production and improving medical care, biotechnology is regarded by most of us as a good thing. But many people are more skeptical of the benefits of science and technology than they used to be, and less trusting of those who control it. Many worry that this powerful technology will be misused or get out of hand. Recent surveys have found that people are almost equally divided in their expectations about biotechnology. Two-thirds believe it offers benefits, yet two-thirds also believe it holds potential dangers (obviously, many people believe both are true).

This book does not pretend to be either all-inclusive or the last word on the subject. But if you are wondering

"What exactly is biotechnology?" or "How should I feel about the applications of biotechnology?," I think you will find this book gives you a thorough introduction to a fascinating subject. In the end, individual attitudes to biotechnology are rooted in our core beliefs about nature and human nature. Whether you believe biotechnology can carry us to an abundant new world or is more likely to leave us floundering on hidden shoals, the outcomes remain to be seen. My own journey of exploration in the research and writing of this book ends with its publication, but I will keep following the story in journals and through the news, as I've no doubt that biotechnology will, in one way or another, increasingly affect my life and the lives of all of us.

I want to thank Trudy Rising for inviting me to write this book and having the faith that I could do it. Michelle Campbell made many encouraging and helpful observations on early drafts, and Susan Lawrence helped to guide the manuscript to its final form. Thorough analyses of the manuscript by Jonathan Bocknek and Dr. Lois Edwards, and a team of reviewers — nonscientists, scientists, proponents and opponents of aspects of biotechnology — all contributed to the book's development, my approach, and my understanding of the differing perspectives that needed to be presented. Naturally, any errors in the text (of which I hope there are few or none), are mine alone.

Eric S. Grace
Victoria, British Columbia

Chapter 1
How Biotechnology Came About

If you want a quick insight into what modern biotechnology is all about, start thinking of yourself as being built and run by molecules. It's thanks to the cooperation of these small chemical units that you and I can blink, breathe, and read. Thanks to molecules, we once grew from microscopic fertilized eggs into functioning human beings.

The amazing thing is that these molecules are nothing special in and of themselves. They are combinations of only half a dozen common elements: carbon, hydrogen, nitrogen, oxygen, phosphorus, and sulfur. Every living thing on the planet is built from the same types of molecules and, at the molecular level of life, every living thing functions in fundamentally the same way, whether a human, a goldfish, a maple tree, or an earthworm.

Biotechnology operates at that molecular level of life, where the seemingly solid boundaries between species disappear. Down among the molecules, there is really no difference between a person and a bacterium. What biotechnology does is choreograph the complex dances among molecules that ultimately make every living thing what it is.

What is biotechnology?

The molecular waltzes of life take place largely inside cells, and one simple definition of biotechnology is "the commercialization of cell biology." More generally, biotechnology is an umbrella term that covers various techniques for using the properties of living things to make products or provide services. The term was first used before the 20th century for such traditional activities as making dairy products, bread, or wine, but none of these would be considered biotechnology in the modern sense. Nor would genetic alteration through selective breeding, or plant cloning by grafting, or the use of microbial products in fermenting. What's new about modern biotechnology is not the principle of using various organisms but the techniques for doing so. These techniques, applied mainly to cells or molecules, make it possible to take advantage of biological processes in very precise ways. Genetic engineering, for example, allows us for the first time to transfer the properties of a single gene from one organism to another. Before I explain these modern techniques in the next chapter, I want to outline some of the history that led to their development.

The thing about biotechnology that surprises most people is that it has produced so many applications so rapidly. Its very pace of development leaves an uneasy feeling of having missed something along the way, as if the whole biotechnology business fell out of the sky fully formed while we were out walking the dog.

It's one thing to be told that scientists can do this or that, another to actually understand how such things came about, to realize how we know what we know. The skills of biotechnology, like all human knowledge, only developed from what came before. The feeling of missing something is the same one you'd get starting a novel halfway through, or catching a TV soap opera for the first time. It won't make much sense if you don't know the story so far. Biotechnology is only the current chapter in a story that began a long time ago.

In the beginning

The path to genetic manipulation can be said to have started in 1665, when the English scientist Robert Hooke published a review of some observations he'd made while peering down a microscope. Describing the tiny spaces surrounded by walls that he saw in samples of cork, Hooke coined the word "cell." He saw similar structures in other plant tissues and supposed their function was to transport substances through the plant.

Ten years after Hooke's publication came out, a Dutch draper and skilful lens grinder named Anton van Leeuwenhoek was making history, designing microscopes with magnifying powers as great as 270 times. Using these instruments, he became the first person to observe and describe microorganisms, which he called "very little animalcules."

Leeuwenhoek accurately calculated the size of bacteria 25 times smaller than red blood cells, and discovered the existence of sperm cells in semen from humans and other animals. Until then, scientists believed that the development of an animal began with the egg, which the mysterious male contribution stimulated to grow. Leeuwenhoek revealed for the first time that fertilization involved both male and female cells equally.

Figure 1.1
Anton van Leeuwenhoek (1632-1723)

He was the first person to see bacteria, using high-quality lenses he ground himself. Specimens were set at the tip of an adjustable screw and illuminated from behind by a candle flame. Crude drawings of Leeuwenhoek's observations show that he was able to see shapes of bacteria that are seen with much more sophisticated microscopes today.

During the 1700s, many other scientists used the new-fangled microscopes to peer into life's hidden dimensions. They found cells throughout every part of both plants and animals, and added more new discoveries to the list of single-celled organisms. But despite seeing cells everywhere they looked, nobody came up with the idea that cells are fundamental to all living things until more than 170 years after cells were first seen — which shows that seeing and understanding don't always go hand in hand.

Two German biologists, Matthias Schleiden and Theodore Schwann, first put that idea into words in 1839, giving us a theory that forms one of the key understandings of biology. The cell theory says that all organisms are made of cells. Some consist of only a single cell; most are collections of many individual cells. (A human body, for example, has an estimated one hundred trillion of them.)

Schleiden and Schwann's conceptual breakthrough seems simple enough, but it has some profound implications. Cells aren't merely soft construction blocks, the basic structural units of life. They are also the basic functional units of life. A single cell is itself alive, potentially carrying out all the processes needed to maintain life within its microscopic space.

Here, for the first time, scientists saw the exciting possibility of finding a tangible answer to the age-old question: "What is life?" If a cell is alive, all the ingredients for making living organisms could eventually be found inside cells. Investigators set off in pursuit of that holy grail during the second half of the 19th century, giving birth to the science of cell physiology.

One debate that took a curiously long time to settle was the question of where cells come from in the first place. The theory of free cell formation persisted well into the 19th century with the proposition that cells can materialize from other substances, much as crystals form in saturated solutions. That idea was vanquished only after improved microscopes and better techniques for preparing specimens let scientists watch living cells grow and divide to make new cells. These observations led them to conclude that cells can only come from previously existing cells.

It doesn't take long to see where that provocative line of thought leads. The concept that all cells come from previously existing cells implies that all life on earth is connected by lines of descent that go back unbroken to one or more original cells — to the very beginning of life on earth! Paradoxically enough, the best evidence for this wasn't initially found by delving into ever-smaller parts of the cell, but by studies of whole organisms in all their glorious variety.

The voyager and the monk

Why are there so many different types of organisms in the world? What makes a particular organism, such as a cat, produce more cats and not, say, dogs? Why do some kittens in a litter look just like their mother while others don't? Answers to questions such as these came from two brilliant and original thinkers of the 19th century: Charles Darwin and Gregor Mendel. These two men set off an explosion of ideas and debates that rumble on today, and are still felt in some of the controversy surrounding biotechnology. They also created two of modern biology's great cornerstones: evolution and genetics.

Charles Darwin was a careful observer who developed his ideas about natural selection during a five-year voyage around the world as naturalist on board the British naval ship *Beagle*. Struck by differences he saw among finches and tortoises on the widely scattered Galapagos Islands in the Pacific, Darwin inferred that isolated populations change as they become adapted to different conditions. Eventually, these changes result in new species with different features. Ancient fossilized remains of now-extinct animals and plants confirm that life on earth has indeed changed dramatically over time — that the species living today are not the same as those living in the past.

A striking coincidence

English naturalist Alfred Russel Wallace proposed a theory of natural selection independently of Charles Darwin. He sent his ideas to Darwin in an essay in 1858. "I never saw a more striking coincidence," Darwin wrote to a colleague.

Darwin's book, *On The Origin of Species by Natural Selection*, appeared in 1859, selling out all 1,250 copies of the first edition on the day it was published. From that time onward, educated people could never again think about life on earth as they might have in the days before Darwin. But the idea of evolution itself didn't originate with Darwin. Various thinkers had tossed that idea around since at least the time of ancient Greece. What Darwin came up with was the means by which evolution takes place.

Darwin's theory makes two important points relevant to biotechnology. First, every species is ultimately related to every other through common ancestors, no matter how much they might differ from one another in appearance today. Second, the theory implies that a record of the evolutionary past is present inside every living thing.

Echoes of the past are most obvious in anatomical remnants, such as the bony vestiges of hind legs found inside the bodies of whales. But the overlap among species is even more profound at the unseen, molecular level, in ways unknown and perhaps not even guessed at in Darwin's time. Nature has been much more conservative in making changes to biologically important molecules than to bodies. In particular, the eventual discovery of the DNA molecule as the agent of heredity in every living organism was both confirmation of Darwin's unifying idea and the basis of biotechnology's great success.

Darwin knew nothing about genes, although the foundations of genetics were being laid during his lifetime by that other groundbreaking scientist, Gregor Mendel. Mendel was an Austrian monk and teacher who began his experiments with plant breeding in a monastery garden in 1856. And even though Mendel founded the science of genetics, he didn't know about genes either.

Mendel discovered the laws of heredity — the statistical relationships that govern how characteristics are passed from one generation to the next. These laws form the basis of evolutionary change, but Mendel was interested mainly in finding out how to predict the outcomes of his crossbreeding experiments.

One of the most important results of Mendel's work was his demonstration that inherited characteristics are determined by discrete factors (which we now call genes) that pass from generation to generation. He also inferred that each organism contains two copies of each factor: one inherited from its mother and one from its father.

These ideas are so familiar to us now it seems they

should have been obvious, but they were difficult to develop. Many elements muddy the actual outcome of breeding in most animals and plants, making it very difficult in practice to predict exactly what the next generation will look like. Darwin, for instance, thought parental characteristics merge in some indefinable way, like mixing different colors of ink. But Mendel had luck as well as genius on his side. The characteristics and the plants he chose to study happen to be uniquely clear and simple in their pattern of inheritance. Mendel's discoveries gave us the concept of the gene as a real, physical presence inside cells. The next step was to find out what that physical presence consists of.

Colored bodies

The way in which organisms reproduce themselves demonstrates the primacy of cells, for every living individual starts life as just a single cell. All the information needed to build the organism must reside in that cell, information given to it by the reproductive cells of its parents, as they were given it by their parents, and so on.

The most likely carriers of information seen inside animal and plant cells were chromosomes, threadlike bodies that make a distinctive appearance in the nucleus of a cell just before it divides in two. The word *chromosomes* means colored bodies. They're called that because they readily absorb the colored stains that scientists use to make cells easier to view under the microscope (Figure 1.2).

Every species has a specific number of chromosomes, consisting of a set of near-identical pairs (one in each pair from one parent, one from the other parent). For

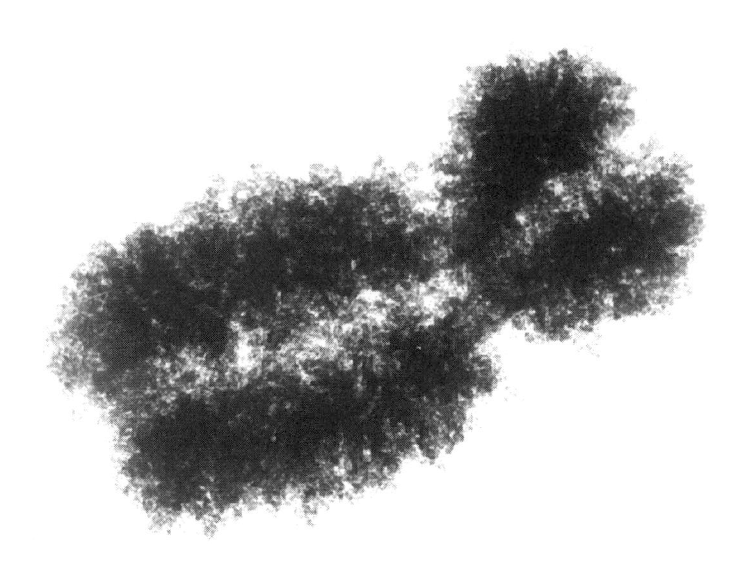

Figure 1.2 An electron micrograph of a human chromosome. Each chromosome is duplicated just before a cell divides, producing two tightly coiled threads (largely DNA) joined together at a narrow constriction. The threads separate at the constriction, and one thread goes to each of the two new cells formed by division, so that each cell of the body has an identical set of chromosomes.

example, humans have 23 pairs of chromosomes, or 46 in total (see table opposite). Every cell of your body has the same number of chromosomes *except* the reproductive cells (eggs or sperm), which have exactly half the usual number. This makes sense given that a new individual life is formed by the fusion of two reproductive cells. When a human egg and sperm combine, the 23 maternal and 23 paternal chromosomes in the reproductive cells add together, giving the fertilized egg equal hereditary information from both parents and keeping the chromosome number constant from one generation to the next.

A correlation between the behavior of chromosomes during fertilization and Mendel's mathematical predic-

Chromosomes in the body cells of different species

organism	number of chromosomes	organism	number of chromosomes
mosquito	6	human	46
fruit fly	8	chimpanzee	48
garden pea	14	potato	48
corn	20	amoeba	50
frog	26	horse	64
earthworm	36		

tions about inheritance led scientists to propose that genes must be located on chromosomes. The first clear evidence for this idea came in 1910, when American geneticist Thomas Hunt Morgan got some unexpected results from crossbreeding fruit flies.

Morgan found a mutant male fly that had white eyes instead of the usual red color. When he used it for further breeding experiments, he found that the inheritance pattern for white eyes precisely followed the inheritance of the X chromosome, one of the two sex chromosomes (X and Y). Morgan realized that his results could be explained if the gene for eye color were actually located on the X chromosome, an example of a characteristic called a sex-linked gene.

Fruit flies in the lab

Fruit flies, *Drosophila melanogaster*, became the standard lab animals used in the study of genetics during the first years of this century. They are ideal for the job because they are easy and cheap to keep, have a life cycle of only two weeks, and contain all their genetic information in only four pairs of chromosomes.

Let there be DNA

In 1869, a young Swiss chemist named Johann Miescher wanted to know what chemicals occur in cell nuclei. He analyzed the material he extracted from white blood cells found in pus and named the substance nuclein. A few years later, he separated a phosphorus-containing acid from his cell substance, and renamed the chemical nucleic acid. Miescher had discovered DNA (deoxyribonucelic acid), the material from which genes are made, but the significance of this molecule wasn't recognized until 75 years later.

There's an irony in the long time lapse between the discovery of DNA and its recognition as genetic material. Scientists had realized that hereditary information could be encoded in large molecules because large molecules are built up from strings of smaller subunits, just as words are built up from strings of letters. But they believed the most likely candidates for this job were big, complex protein molecules. DNA was assumed for many years to be both too small and too simple to hold the vast amounts of detail needed for the instructions to build new organisms.

In 1928, British scientist Fred Griffith carried out an experiment that set researchers on the right track. It involved two strains of bacteria: a virulent strain that causes pneumonia and a mutant, harmless strain. When Griffith injected mice with either the harmless strain, or a preparation of the virulent strain that had previously been killed by heat, the mice suffered no ill effects. But when he injected the harmless strain together with the heat-killed lethal strain, most of the mice died within two days (Figure 1.3). When he examined their blood, Griffith found live, virulent bacteria in it!

Figure 1.3

The Griffith experiment

Griffith's experiment showed that genetic material could move from one strain of bacteria to another. His work eventually led other researchers to identify DNA as the material from which genes are made.

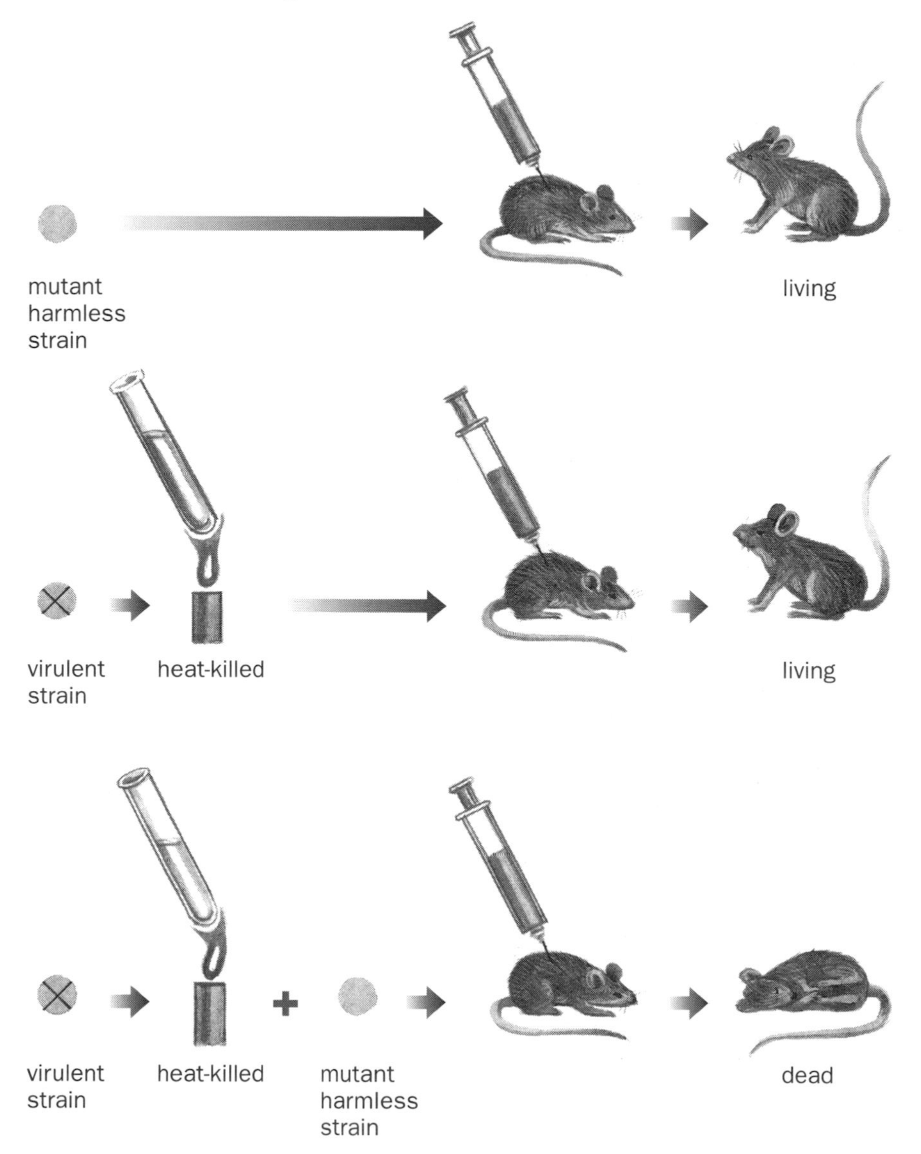

Mystified at first by the resurrection of the lethal bacteria, scientists eventually suspected that the virulent property was somehow passed from the dead bacteria to the living and previously harmless strain by a "transforming principle." The transforming principle was, in essence, genetic material, carrying the trait of virulence from dead cells to living ones. If the transforming principle could be extracted and isolated, scientists would at last know what genes are made from.

The second half of the puzzle was finally solved by Oswald Avery and his coworkers in New York in 1944. They spent years grinding up bacteria, refining and purifying their extracts, and adding chemicals until everything was eliminated but the one essential transforming principle. What they ended up with was DNA. It must be DNA, they concluded, that carries hereditary information.

Unraveling the double helix

There was great excitement among molecular biologists during the 1930s and '40s. The physical basis of heredity was rapidly becoming better understood, and scientists felt they were close to peering inside the hidden machinery of the cells, into the "little black box" that directs what each living thing is to become. They knew that:

- heredity is controlled by discrete factors called genes
- genes are located on threadlike chromosomes found in cell nuclei
- genes are made from DNA.

Clearly this DNA molecule needed looking into. Investigators were helped in their quest by powerful new tools and techniques designed to analyze atomic structure. The search for an answer to the question "what is life?" had by now subtly but significantly shifted — from cells to chemicals. This concept wasn't new, however. In the late 1700s, the founder of modern chemistry, Antoine-Laurent Lavoisier, had said, with either great perception or great presumption: "*La vie est une fonction chimique*" (Life is a chemical process).

Analysis had already told scientists the chemical composition of DNA. Its building blocks are sugar, phosphate, and four different nitrogen-containing bases named guanine, cytosine, thymine, and adenine (shortened for convenience to G, C, T, and A). But the important question remained: How are these smaller molecules linked together in the larger DNA molecule? The answer to that, it seemed, would also tell scientists how DNA is able to store and pass on a practically infinite number of bits of hereditary information.

The answer came in 1953 from James Watson and Francis Crick, two young researchers at Cambridge University. They built a model of a molecule as big as themselves from pieces of bent wire and brass, and tinkered with it until they found the structure that best fitted everything then known about DNA. It was a fairly simple spiral coil of two linked strands — the now-famous double helix.

Figure 1.4

DNA molecule

How to put together a DNA molecule from sugar, phosphate, and four bases: guanine, cytosine, thymine, and adenine. The last two twists show different stylized ways of representing bonding.

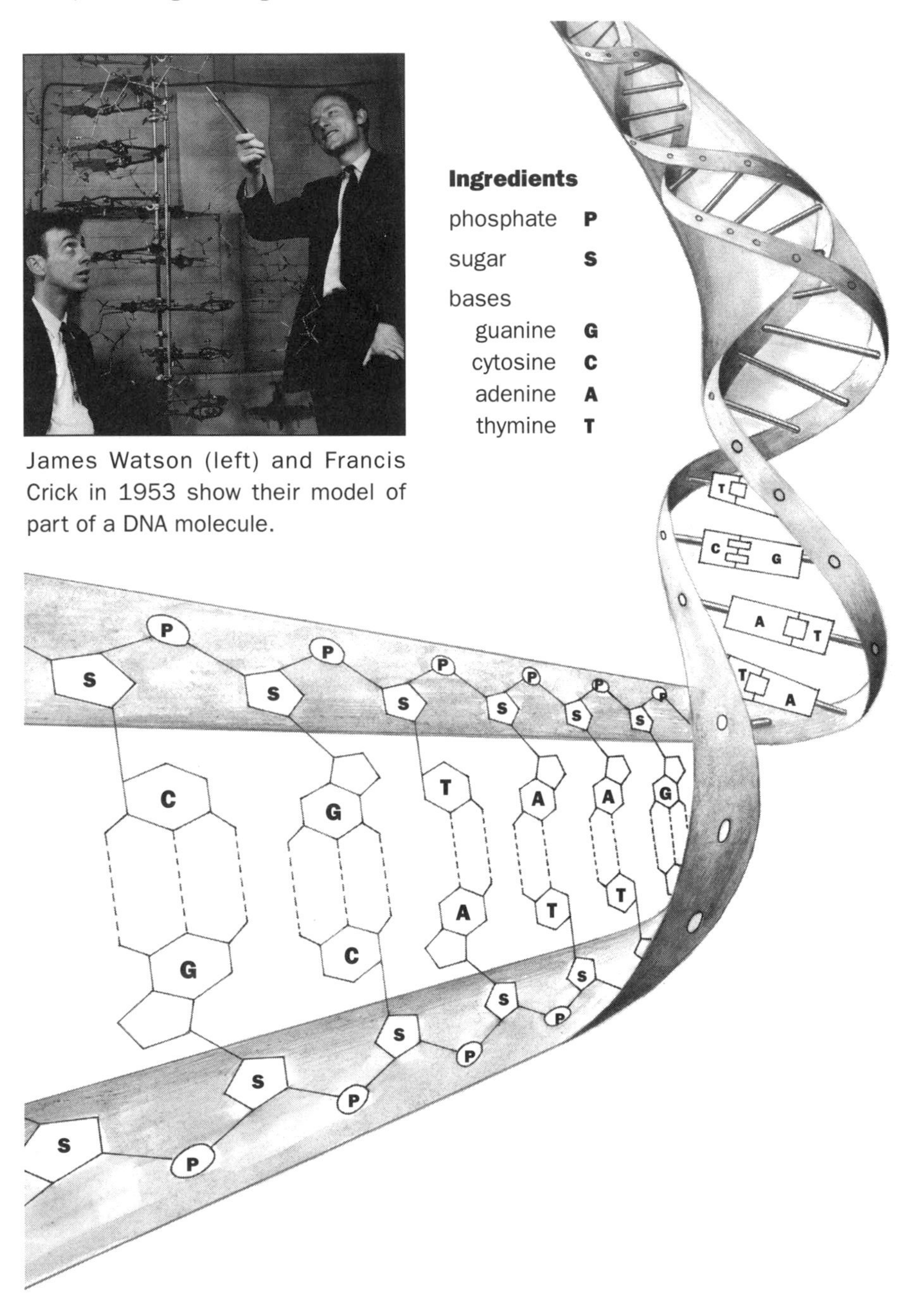

James Watson (left) and Francis Crick in 1953 show their model of part of a DNA molecule.

The rails of this graceful winding staircase are alternating molecules of sugar and phosphate. The staircase steps that join the two rails together are pairs of bases (Figure 1.4). With the discovery of this elegant pattern, history changed gears and the modern era of genetic manipulation was on its way.

How does DNA store information?

The key to DNA's astonishing power to store information lies in the four different bases (G, C, T, and A). They form the letters of the genetic alphabet. Imagine yourself walking up a DNA molecule on one side of the steps, reading off the bases as you go. Your journey might read AGGTCTATCAGC, and so on. Another section of steps further along will give you a completely different sequence of letters. In fact, the four bases can be arranged along the DNA molecule in a practically infinite variety of sequences. A given sequence spells out a given gene. Different genes have different sequences and different lengths (numbers of bases). That's really all there is to it.

A copy in every cell

All the cells in your body have essentially identical copies of the unique DNA sequences that were put together at the moment of your conception. To get copies from that original single cell into every cell of the body, DNA molecules must faithfully duplicate themselves each time a cell divides in two. What makes this possible is the way the bases on the two strands of the double helix complement one another.

If you take another look at the double helix in Figure 1.4, you'll see that the four bases form only two types of pairs. An A base always joins to a T base, and a G always joins to a C. Whatever the sequence of bases along one strand, therefore, the sequence along the other strand always complements it in a predictable way.

To duplicate itself, a DNA molecule simply splits apart down the middle and then rebuilds matching parts to each half (Figure 1.5). The two separated, single strands of DNA use their own base sequences as templates to reconstruct their other halves, making two identical copies from the original one.

This answers the question of how DNA passes on genetic information from cell to cell. But what exactly is that genetic information?

What do genes do?

To say that a gene consists of a particular sequence of bases in a DNA molecule isn't a very satisfying description. It doesn't tell us anything about what bases actually do, or how they do it. To the average person, a gene is something that gives you, say, blue eyes or brown eyes. So how does a sequence of bases in a DNA molecule do that?

To get a clue, let's turn to British physician Archibald Garrod and a few of his patients. In 1902, Garrod was examining a disorder named alkaptonuria, in which the patients' urine turns black when exposed to air, due to a particular acid in the urine. The disease was known to

Figure 1.5

DNA replication

A DNA molecule produces two identical copies of itself.

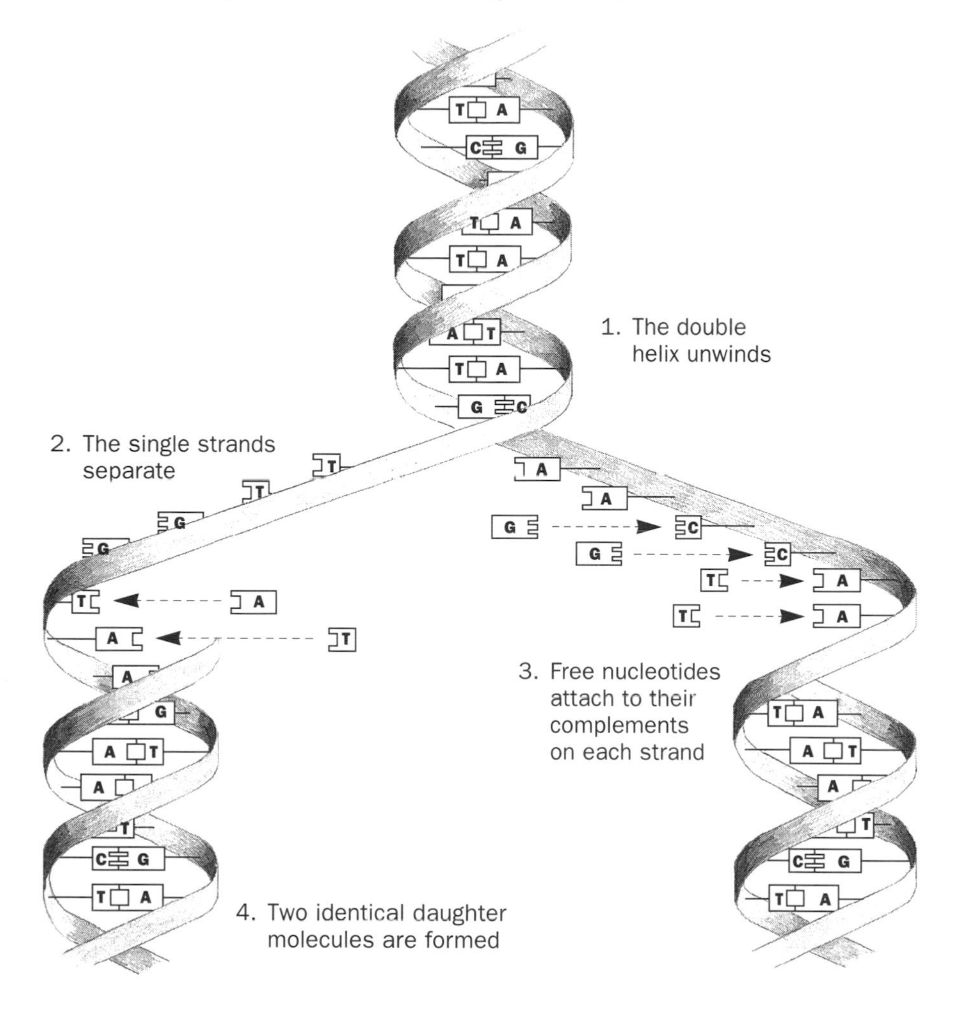

run in certain families for several generations so it is clearly inherited and, therefore, controlled by genes.

In normal people, the acid responsible for the urine's black color is broken down in the body by a chemical reaction. Garrod logically concluded that his patients lacked something needed for this reaction. Specifically, they lacked a certain enzyme — a protein that acts as a biological catalyst, allowing chemical reactions to take place rapidly at body temperatures.

Since the disease is genetic in origin, Garrod speculated that genes consist of instructions for making enzymes, and perhaps other types of proteins as well. His insight was right, but it wasn't confirmed until a series of clever experiments gave definite proof nearly 40 years later.

In 1941, Stanford University geneticists George Beadle and Edward Tatum made the breakthrough that indisputably tied genes to enzymes. They did this through a series of tests with genetically mutated strains of bread mold. Each strain lacked the ability to produce one of the essential nutrients (amino acids or vitamins) that fungi normally need to grow. This lack, in turn, was due to the absence of a necessary enzyme.

By growing different strains of bread mold on different dishes with different combinations of nutrients, the scientists determined exactly which particular enzyme was lacking in each mutant strain. At the same time, they found that each genetic mutation was located at a specific site on the fungal chromosomes. A different site (or, in other words, a different gene) was associated with each enzyme. The geneticists concluded that one gene produces one enzyme.

So the answer to the question "what do genes do?" is that genes are instructions for making various proteins. Who can believe that the difference between blue eyes and brown eyes, or for that matter between a sheep and your next-door neighbor, comes down to that?

Protein primer

We return now to the idea I began with at the start of this chapter — the notion that people and all other organisms are built and run by molecules. It isn't a metaphorical or a metaphysical idea — it's literally true. To convince you of this, I'll introduce you to protein molecules.

Protein is one of the things listed on my boxes of breakfast cereal, but to a molecular biologist, proteins are the very foundation of living systems. Virtually every process and product in living cells depends on proteins. They do everything from activating essential chemical reactions, to carrying messages between cells, to fighting infections, to making cell membranes, tendons, muscles, blood, bone, and other structural materials.

Examples of proteins

For structure	For function
collagen (found in bone and skin)	hormones (control body functions)
keratin (makes hair and nails)	antibodies (fight infection)
fibrin (helps clot blood)	enzymes (help speed up chemical reactions in the body)
elastin (major part of ligaments)	hemoglobin (carries oxygen in the blood)

Since proteins are responsible for practically all of a cell's distinctive properties, we can say that proteins make an organism what it is. Proteins make the differences between, say, a hormone-secreting cell in your pancreas, a muscle cell in your biceps, a nerve cell in your eye, and a bone cell in your rib. Proteins make the differences between the hair on your head, the wool on a sheep, the feathers on a sparrow, and the scales on a goldfish.

Proteins are us

The human body alone contains over 30,000 distinct types of protein, each having its specific uses. Other organisms have some of the same proteins, as well as different proteins not found in humans. Enzymes are the biggest single class of proteins. An average mammalian cell contains about 3,000 enzymes.

Despite their many different functions, all protein molecules are constructed in the same basic way. They are long, folded chains of smaller molecules called amino acids. There are 20 different types of amino acids in all, which can be combined in an almost infinite number of ways to produce different proteins. One of the smallest proteins, insulin, contains more than 50 amino acids, while most are very much bigger, typically containing from a few hundred to over a thousand individual amino acids (Figure 1.6).

The amino acids

You may have seen some of the names of these 20 amino acids listed on the labels of food products.

arginine, asparagine, aspartic acid, cysteine, glutamic acid, glutamine, glycine, histidine, isoleucine, leucine, lysine, methionine, phenylalanine, proline, serine, threonine, tryptophan, tyrosine, valine.

Figure 1.6 Protein molecules are made up of long chains of amino acids that twist, coil, and often fold on themselves. Each sphere shown here represents a single amino acid.

The numbers, types, and arrangement of amino acids in a protein molecule determine its structure, and its structure determines the job it will do in a living organism. The shape of some proteins is very sensitive to the arrangement of particular amino acids, and a change in the identity of only one amino acid can cause very subtle, or very profound, effects — like a misspelled word altering the meaning of a sentence.

Some diseases, like alkaptonuria (which I described on page 18), are the result of badly made or missing protein molecules, produced by a genetic mutation. This sort of disorder might be prevented by gene therapy, a subject explored in Chapter 3.

Genes, proteins, and your eyes

The color of your eyes is determined by the set of genes you got from your parents. But in more immediate and practical terms, the color is determined by the actions of proteins (enzymes). Eye color is the result of the amount and distribution of a pigment in your iris. The production and deposition of the pigment depends on a series of chemical reactions involving enzymes. People with blue eyes, for example, lack the enzymes needed to deposit pigment in the iris. With a different set of genes, you might have different enzyme action, a different deposition pattern of pigment, and a different eye color.

The three-dimensional shapes of protein molecules, with their complex twists, turns, and folds, can make their impact felt on the larger world of our senses in some remarkably direct ways. The physical properties of these molecules are echoed in the properties of the materials they form. For example, the molecules of keratin that make up hair and muscles are spiral and springy. Collagen molecules, found in bone, skin, and tendon, are ropelike in structure. The protein molecules in silk fibers have a smooth, sheetlike shape.

We can change the shapes of some protein molecules, and their properties, by heating them, adding chemicals, or even by simple physical means. Take ketchup, for example. In a bottle on the shelf, ketchup's long protein molecules are coiled together like clumps of cold spaghetti, making the ketchup thick and sluggish. When you shake the bottle, you break the molecules apart and loosen them up. With separated molecules, the shaken ketchup flows more readily. Egg white proteins do the opposite. As you beat an egg white, its long-chain proteins get more tightly tangled and bound together, slowly transforming the substance from a thick, semitransparent liquid to a stiff, white foam.

I hope this brief taste of protein chemistry gives you some idea of the vital roles proteins play in all living systems. The power of biotechnology comes from knowing how particular proteins are built from particular genetic instructions.

The genetic code

An important thing to notice about the structure of DNA is that it is built up by repeating subunits of three linked molecules: base, sugar, and phosphate (Figure 1.7). These units, named *nucleotides*, are the fundamental components of DNA. Because there are four different bases, there are four different kinds of nucleotides.

Figure 1.7
Nucleotides

A strand of DNA is a polynucleotide: a long chain of nucleotides, connected to one another by chemical bonds.

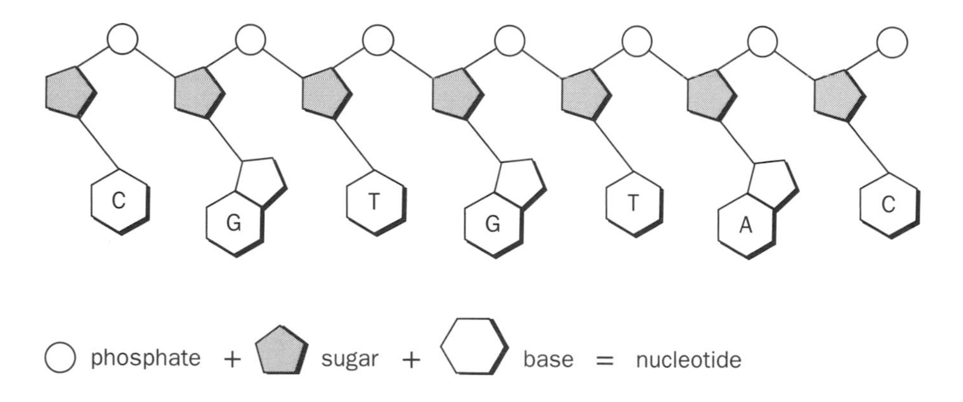

The question is, how do four different nucleotides translate into 20 amino acids and thousands of different proteins? It's not a big problem, really, just a matter of coding. Think of the dots and dashes of the Morse Code giving alphabetic instructions for writing out *King Lear* and you'll have an idea of how it can be done.

Like the dot-dot-dot and dash-dash-dash signifying "S" and "O" in Morse code, the genetic code is organized in groups of three. That is, a sequence of three adjacent nucleotides is a code for each amino acid. For example, the amino acid glycine is coded by the sequence GGA. Each triplet of nucleotides is called a codon. Since four nucleotides can be put together in groups of three in 64 different ways (4 x 4 x 4 = 64), there are more than enough codons to encode all 20 amino acids. (In fact, most amino acids are encoded by more than one codon.)

The complete genetic code linking each of the 64 codons to an amino acid was finally cracked by the research of Har Gobind Khorana and Marshall Nirenberg in 1967. As a result of their work, they were able to draw up a universal decoder chart showing the correlations between codons and amino acids — a correlation identical in virtually all organisms.

I said before that genes are instructions for making proteins. I can now refine that definition and say that a gene is a segment of DNA with a unique sequence of nucleotides, encoding information for assembling particular amino acids into a particular protein.

Figure 1.8

Nature, the expert packer

Recall the chromosomes. Each chromosome is about 40% DNA, consisting of a very long double helix, tightly wound and coiled around protein cores and extending unbroken through the entire length of the chromosome. A DNA fiber in a typical human chromosome has about half a billion nucleotides, giving it an astronomical number of possible sequences.

If you could take the strand of DNA from one chromosome and lay it in a straight line, it would measure about 5 cm (2 in) long. With 46 chromosomes in a human cell, this means that the DNA content of one cell, stretched out and laid end to end, would be over 2 m (6 ft) in length! How much DNA is there in a human body? Enough to reach from here to the sun and back about 500 times. Nature is clearly an expert at packing.

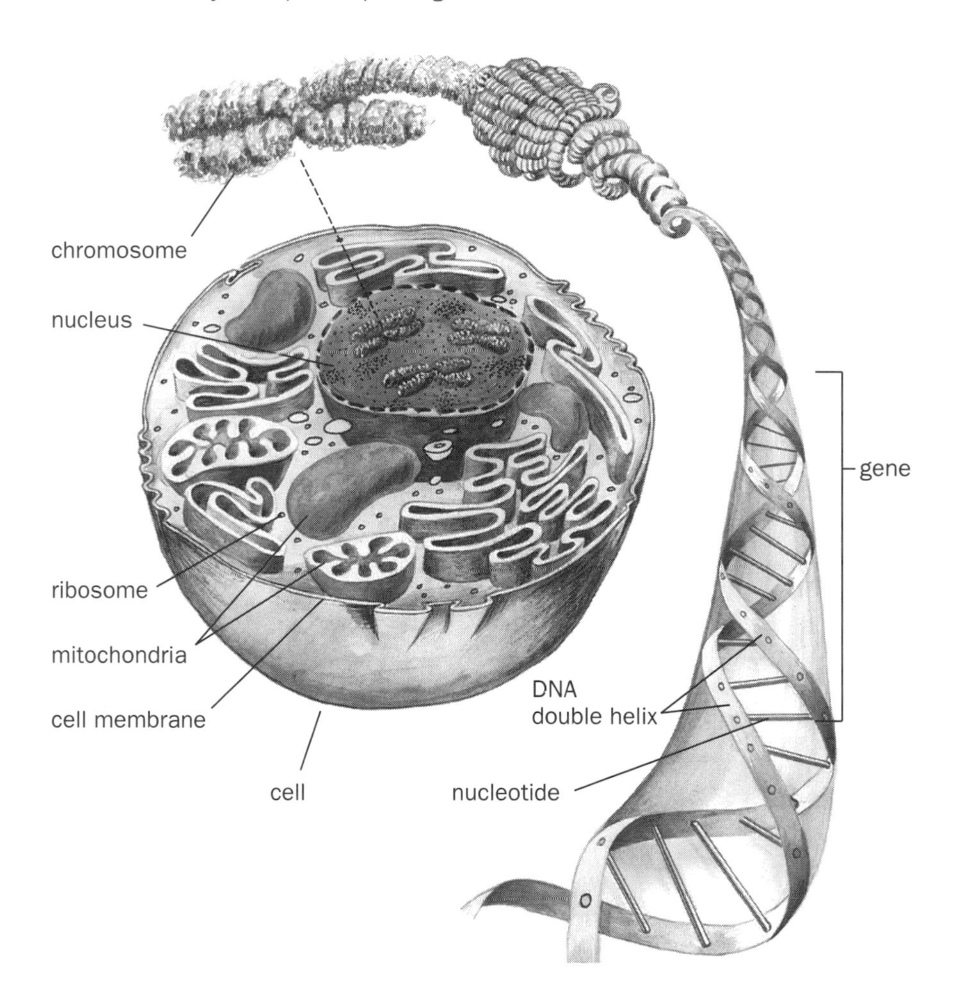

How genes make proteins

It becomes difficult at this point in the story to avoid a textbook-like complexity, with descriptions of sections of genes that don't code for amino acids, sections that carry instructions for starting and stopping protein building, sections that overlap with each other, and so on. But although there's much more that can be said about the action of genes, you don't need these additional details in order to make sense of biotechnology. The table below summarizes the history outlined so far. Before bringing this account to an end, I'll just make one more point about the link between genes and proteins.

Steps on the road to biotechnology

Year	Event
1665	Robert Hooke describes and names cells
1675	Anton van Leeuwenhoek develops better microscopes and discovers microorganisms, bacteria, and sperm cells
1839	Matthias Schleiden and Theodore Schwann state their cell theory
1859	Charles Darwin publishes *On The Origin of Species*, establishing the theory of natural selection
1866	Gregor Mendel publishes *Experiments With Plant Hybrids*, outlining the principles of heredity
1869	Johann Miescher makes the first chemical analysis of nucleic acid
1902	Archibald Garrod speculates that genes consist of instructions for making proteins
1910	Thomas Hunt Morgan establishes that genes are located on chromosomes
1928	Fred Griffith finds that a "transforming principle" (genetic material) carries the trait of virulence from dead bacterial cells to live ones
1941	George Beadle and Edward Tatum establish that one gene makes one enzyme

1944	Oswald Avery and his team prove that Griffith's "transforming principle" is DNA
1953	James Watson and Francis Crick deduce the structure of the DNA molecule — a double helix
1967	Har Gobind Khorana and Marshall Nirenberg crack the genetic code

DNA is the storage vehicle for genetic information, but it doesn't directly do the work of building proteins itself. DNA is the boss. The workhorse is a similar nucleic acid, RNA (ribonucleic acid), which carries out DNA's instructions. Essentially, RNA assembles proteins, one amino acid at a time, using the sequence of nucleotides along a strand of DNA (that is, a gene) as its guide.

A protein molecule is made by a gene in two stages. First, an RNA copy of the gene is made. Transcribed from a template of DNA, the RNA copy has a nucleotide sequence complementing that of the gene. Then the RNA moves to another part of the cell, where its nucleotide sequence is translated into a sequence of amino acids to build a protein.

The cell's tiny protein-assembly plant works in much the same way in all organisms. That is why genetic engineers can take genetic instructions from one organism and add them to another, or even write their own new instructions.

D(aring) N(ucleotide) A(dventures)

Outside of its cell, there is no distinction between a human gene, a cat gene, a wheat gene, or a bacterial gene. There is nothing intrinsic to a gene, in other words, that tells you what species it comes from. The function of a gene is to produce a protein: nothing more, nothing less. The differences between species, or between individual organisms, lie only in the particular numbers and specifications and combinations of their genes.

Because the genetic code is universal, almost any cell in any organism can "read" a gene and translate it into the relevant protein. Today, for example, the insulin used to treat thousands of people with diabetes is produced on an industrial scale in huge vats by bacteria that have been genetically engineered to carry the human insulin gene. This is the essence of biotechnology.

With an understanding of how cell mechanisms produce proteins, our journey which began with the discovery of cells themselves comes to rest. It took several generations of scientists to show us that:

- the properties of living things come from the properties of the proteins they contain
- the properties of proteins depend on the arrangement of amino acids making them up
- the arrangement of amino acids is determined by the sequence of nucleotides on a section of DNA — or in other words, a gene.

This has been a story about how we discovered some of nature's secrets and learned how organisms come to be the way they are. Biotechnology is using those secrets to alter organisms in very specific ways, and how it does that is the business of the next chapter.

Chapter 2
Tools in the Genetic Engineering Workshop

I asked the woman at the biotech trade display what the little microwave-size machine did. "It's a PCR machine," she said, beginning to demonstrate. She opened a lid to expose a grid of small wells in the heart of the machine. "Put your samples in there, close the lid, punch in the numbers, and wait."

So simple even a child could do it, provided the child has the $7,000 currently needed to buy the machine. This is now all it takes to clone DNA. (Cloning simply means making multiple identical copies, whether of genes, molecules, cells, or whole organisms.) Starting with a sample of as little as a single fragment of DNA, the automated process can produce a million identical fragments in only a couple of hours; you can get a billion if you wait another hour.

The woman had no doubt that sensitive and accurate DNA copiers like this have an assured future. "They'll soon be in doctor's offices to test for diseases," she said.

"You only need one cell, a drop of blood, one sperm. Then you can copy it and run multiple tests."

You can detect some viral diseases this way a year or more before symptoms show up. Or take a few fetal cells from a pregnant woman's blood and carry out a genetic analysis without having to disturb the fetus. "They use these in court cases," she pointed out. And indeed PCR machines are already widely used in forensics for amplifying minute samples — say, the saliva from the back of a postage stamp — and constructing DNA "fingerprints" that match individuals to the scene of a crime. Historians can use PCR technology to study evolution and past diseases, using fragments of DNA from mummies or other relics. Food inspectors can take samples of hamburger or sausage and find out what animal meats were used in them (any horse in there?), or discover from a drop of wine exactly which grapes went into the bottle.

PCR stands for polymerase chain reaction. It's one of the common tools of biotechnology I'll be describing in this chapter, along with DNA "fingerprinting," gene probes, recombinant DNA, cloning, and a number of others that will come up later in this book. All in all, most of these techniques boil down to ways of cutting up DNA molecules, locating particular genes, joining one length of DNA to another, duplicating genes, and modifying organisms by introducing new genes into them.

Many of the practical techniques of biotechnology are based on the naturally occurring properties of various cells, genes, and enzymes, which researchers have discovered and then adapted for their own purposes. Much of the ground-breaking research that led to the development of genetic engineering focused on bacteria, and these omnipresent microbes (often thought of only in connection with diseases and other unpleasant facts of life) still play a central role in many applications of biotechnology.

Why bacteria?

Anyone who wants to study bacteria needs only a small spoonful of garden soil or a scraping from inside somebody's mouth to get about 10 million subjects for investigation. The most numerous of single-celled organisms on earth, bacteria are easy to house, cheap to feed, and multiply rapidly — to say the least. Given the right conditions, a bacterial culture may double in weight in as little as 20 minutes. On a commercial scale, this growth rate lets managers quickly clone genetically altered bacteria and put them to work making hormones, antiviral compounds, enzymes, vaccines, and other valuable products. In the lab, the rapid turnover of generations gives research scientists plenty of data for analysis of genetic change.

More significant even than their numbers is the sheer diversity of bacteria. Over 10,000 different species are spread throughout practically every environment on the planet — in soil and water, on the ocean floor and mountaintops, in ice and hot mineral springs, and on and inside every larger organism.

Their ubiquity reflects their astonishingly complex and variable metabolism, or internal chemistry. Many species of bacteria cannot be told apart by their appearance, but only by the chemical transformations they produce in their cells. To put the diversity of the bacterial world in perspective: your cells are more similar to those of a potato and a shark than one species of bacteria may be to another. This metabolic variety is another reason why bacteria are so interesting to biotechnologists.

A head start

Bacteria are diverse because they have been on the planet for billions of years, giving different species plenty of time to adapt to almost all of earth's varied environments and opportunities. After the first bacteria appeared about 3.5 billion years ago, bacteria had a monopoly on the planet for the next two billion years before other types of single-celled organisms evolved. There were no multicelled organisms until a mere 700 million years ago, which means that single-celled creatures were the only kind of life there was for 80 percent of the time there has been life on earth. Given this head start, it's not so surprising to find that bacteria can live just about anywhere and feed on just about anything, including rocks, oil, plastics, and wood.

A major difference between bacteria and all other forms of life is the way DNA is organized within their cells. In plants, animals, and microorganisms other than bacteria, most DNA is found on chromosomes inside a cell nucleus. Bacteria, however, do not have cell nuclei. They are called *prokaryotes* from the Greek words meaning "before nuclei." (All other organisms are called *eukaryotes*, meaning "with true nuclei.") Bacterial DNA is found on a single chromosome in the shape of a large closed loop. Many bacterial cells also include a few much smaller, independent circles of DNA called plasmids. These freewheeling circles of genetic material can readily pass from one cell into another, giving scientists a vital tool for transferring genes between species.

Gene transfer is something that occurs normally among bacteria, being carried out by a number of different processes. In one of these, called transformation, DNA is released from bacterial cells into the surrounding medium, then taken up and incorporated into the DNA of nearby cells. (The "transforming principle" discovered by Fred Griffith in 1928 and described on pages 12 to 14 was an example of transformation that eventually led scientists to realize that genes are made of DNA.) Another method of DNA transfer involves viruses, which can combine fragments of bacterial DNA with their own. They carry the foreign DNA from one species of bacteria to another when they infect more cells. Researchers have exploited the strategies used in battles between viruses and bacteria to develop a method for making recombinant DNA (that is, novel DNA made by combining DNA fragments from different sources).

Hijackers and molecule snippers

Enigmatic entities occupying the borderlines between living and non-living things, viruses are little more than maverick molecules of DNA or RNA housed in protective protein coats. They resemble cells in having genetic instructions for making new versions of themselves, but differ from living organisms in lacking the biochemical machinery needed for their own multiplication.

Left to themselves, viruses do nothing. They can remain unchanging for years, inert as a jar full of pebbles. To reproduce, they must hijack the metabolic apparatus of a living cell, subverting it to manufacture new viruses and often killing the cell in the process.

Viruses that commandeer bacterial cells are named bacteriophages, or simply phages. They settle on bacterial hosts and inject their DNA strands, leaving their empty protein coats behind on the surface of the bacteria like abandoned lunar modules, now of no further use.

Inside the host bacterium, a virus is at first just a naked strip of DNA. Details coded in this invading DNA direct the bacterial cell to build more viral parts: both more DNA and more protein coats. Eventually, newly constructed viruses of the next generation rupture the cell and disperse into their surroundings (Figure 2.1).

With the knack of modifying the cells of other species to carry out their genetic instructions, viruses can be thought of as the first genetic engineers. Scientists co-opted these skills in some of the first experiments with recombinant DNA, using phages as Trojan horses to smuggle the recombinant DNA into bacterial cells.

The first requirement for making recombinant DNA is to create small DNA fragments. Scientists don't want the entire molecule, after all: only a restricted part holding the genes they're interested in. In the early days of research, DNA molecules were commonly broken into fragments by vibrating them with high-frequency sound waves, but the fragments produced in this way were broken at random and of assorted sizes. In 1970, scientists discovered that bacteria could supply them with far better tools for the job.

The ideal DNA snippers are a group of enzymes called restriction endonucleases. These enzymes probably evolved in bacteria as a defense against viruses. Since viruses attack by sending their DNA or RNA into a cell, bacteria counterattack by chopping up the foreign molecules into bits with their restriction enzymes, restricting the infection.

Figure 2.1

Bacteriophage reproduction

A virus uses a bacterial cell to replicate its genes.

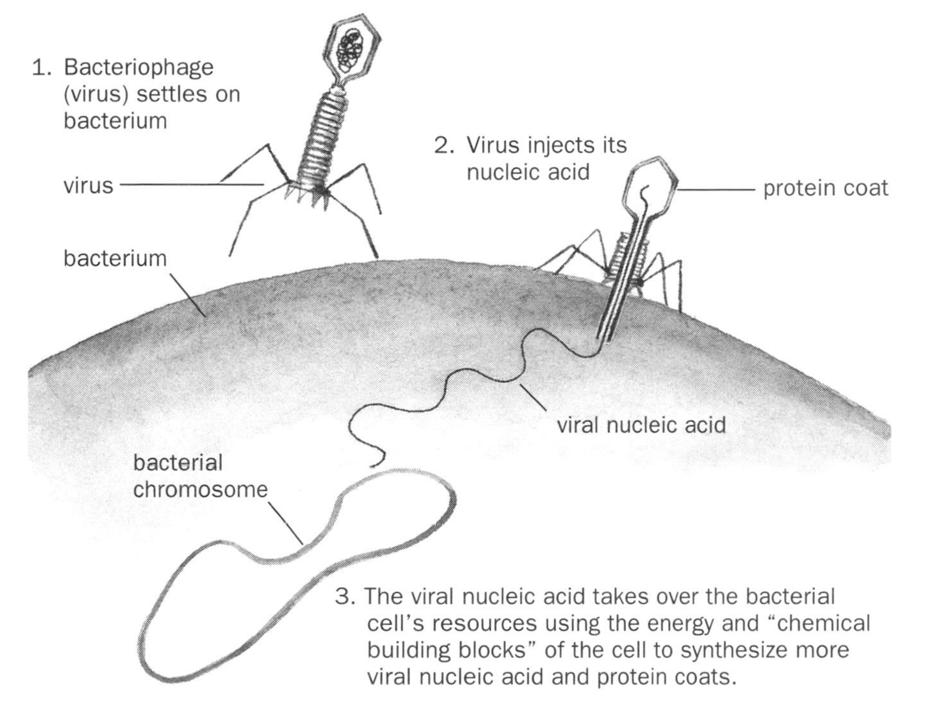

Crucially, each restriction enzyme snips a DNA molecule at specific points only, identified by a particular sequence of nucleotides. Different enzymes recognize and cut different sequences. To date, researchers have discovered more than 800 restriction enzymes, many of which are now routinely produced by commercial companies for use by researchers and manufacturers of "bioproducts."

With restriction enzymes as their cutting tools, not only can scientists produce standard fragments of DNA, they also know that every cut length ends with a particular nucleotide sequence — a fact that later helps them join different fragments of DNA together.

First, catch your DNA

The Victorian cookbook writer Mrs. Beaton begins her recipe for jugged hare with the instruction: "First catch your hare." Before scientists can begin to make recombinant DNA, they need some fairly pure strands of the molecule to work with. Here's one tried-and-true recipe that should give satisfying results every time.

DNA in a tube

Use this DNA in an imaginative way of your choice, either on its own or combined with other ingredients.

Ingredients

chemical broth
bacteria
ethylenediamine tetra-acetate (EDTA)
sodium dodecyl sulfate (SDS)
phenol (carbolic acid)

Method

1. Take a large vat of nutritious chemicals, warm to body temperature, and add a handful of bacteria (about a thousand million). Let stand until the soup is cloudy, with a good density of bacteria.
2. Transfer a small quantity of the mixture to a test tube and spin in a centrifuge at about 8,000 revolutions per minute for 30 minutes. The tube should now have a small, grayish-yellow pellet of bacterial cells at the bottom and a clear liquid above.
3. Pour off most of the liquid and discard. Vibrate the pellet and remaining liquid in a mechanical shaker to separate the cells.
4. Add a drop or two of EDTA and SDS. Between them, these two chemicals break open the bacterial cell walls and release all the cell contents. (EDTA weakens the cell walls by removing their magnesium ions while SDS is a detergent-like chemical that dissolves the fat molecules in the cell walls.) After about 30 minutes you should have a clear, viscous liquid the consistency of egg white.

---

5. Your sticky mixture is made up of long, tangled strands of DNA molecules combined with proteins and a few other bits and pieces from inside the cells. To get rid of the proteins, add a little phenol (carbolic acid) and rock the tube gently back and forth. This mixes in the phenol without breaking up the DNA strands. The proteins will separate from the solution and sink slowly to the bottom of your tube, forming a thick, gray sludge.
6. Spin the sludge in the centrifuge again for 30 minutes. You will end up with a small, clear layer of phenol at the bottom of the tube; a dirty white band of proteins above that; and a clear liquid containing DNA at the top.
7. Carefully remove your DNA a drop at a time, using a fine pipette with a bent tip, and transfer it to a clean test tube.

Figure 2.2 A tube of DNA. DNA released from cells appears as sticky, milky-white strands.

Making recombinant DNA

The snips made by restriction enzymes at a given nucleotide sequence are usually offset on the two strands of DNA rather than directly opposite each other (see Figure 2.3). This leaves both cut fragments of DNA with dangling, single-stranded "tails" of unpaired bases, which are used to bond them to other fragments.

Any two fragments of DNA sheared by the same restriction enzyme can be joined together, since they will have complementary sequences on their dangling strands (often called "sticky ends"). This is true no matter what

Figure 2.3

Restriction enzyme snipping

Restriction enzymes cut both strands of DNA at a specific sequence, leaving "sticky ends" for rejoining to new DNA

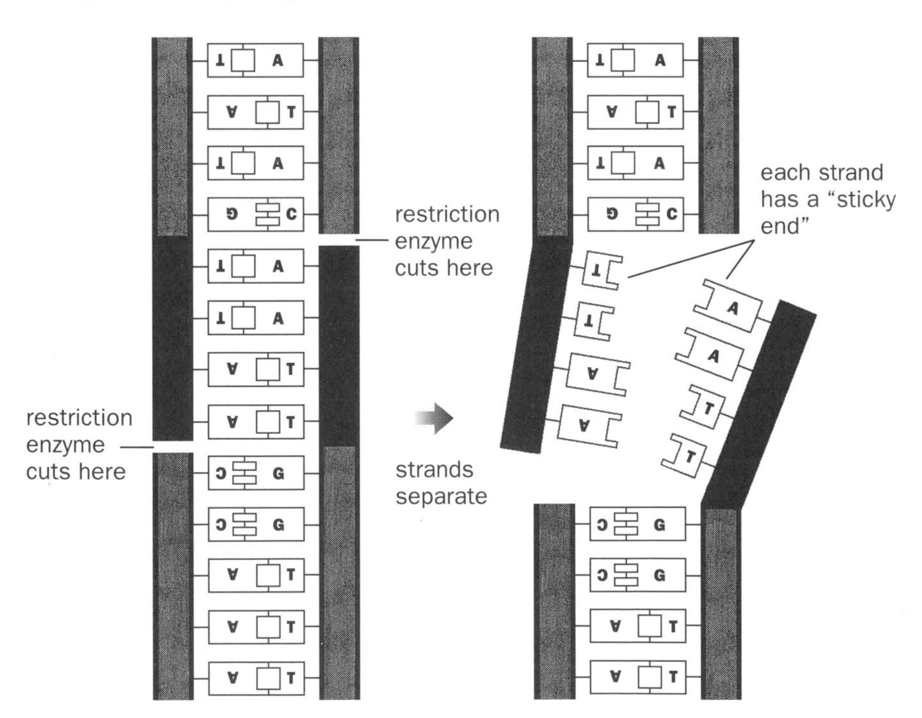

the source of the DNA. So a fragment sheared from the DNA of a mouse, for example, can be joined to a fragment cut from the DNA of an elephant by the same restriction enzyme.

Compatible fragments of DNA bond together when complementary bases on their sticky ends pair up. These bonds, however, are fairly weak and can easily be broken by such things as heat. The connection is made more secure with the help of a sealing enzyme called a ligase. Ligases are another group of naturally occurring enzymes. They are produced by cells to help synthesize DNA and repair minor damage to the molecule.

Putting new genes into cells

The usual reason for making recombinant DNA is to introduce a new sequence into a species where it doesn't normally occur. The added sequence includes a gene that modifies the host cell in some way. For example, scientists may want to introduce a gene for producing an antiviral compound into bacteria, so the bacteria will manufacture this compound for medical use. The challenge is to get the recombinant DNA into the host cells without seriously disrupting their normal functioning.

This is where the plasmids and bacteriophages mentioned earlier come into play. All a researcher has to do is splice the DNA of interest into the DNA of one of these naturally occurring vectors (transmitting agents), then release the vectors with their recombinant DNA in a culture of bacteria and let them do the rest.

The very first transfer of a gene from one organism to another was carried out in this way. In 1973, American geneticists Herbert Boyer and Stanley Cohen used restriction enzymes to cut up large bacterial plasmids, and from the resulting fragments they separated those containing a gene for resistance to an antibiotic. They used the same restriction enzyme to cut up DNA from an African clawed toad, then mixed the toad DNA fragments and the gene-containing plasmid fragments together.

After allowing time for the cut-up fragments to recombine, they added bacterial cells to the mix. They later isolated bacterial cells showing resistance to the antibiotic — those, in other words, that had taken up the plasmid fragments. On further examination, they found that some of these bacterial cells also contained toad DNA, joined into the plasmid loop. Thus, they had produced bacterial cells incorporating toad genes. This type of procedure is now standard practice for making genetically engineered bacteria (Figure 2.5).

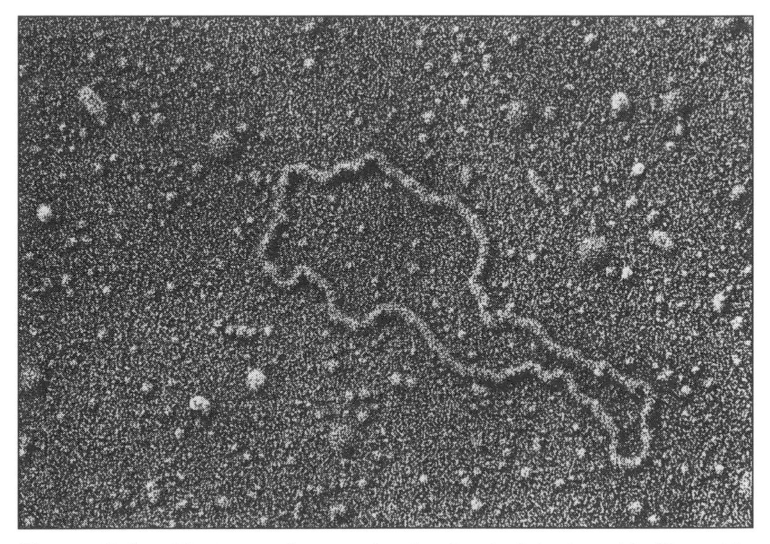

Figure 2.4 Electron micrograph of a bacterial plasmid. Plasmids like this typically have about 5,000 base pairs — enough to code for about five average-sized proteins. Compare that with a human cell's three billion base pairs.

Figure 2.5

DNA with plasmid vector

A gene for human interferon is spliced into a plasmid, introduced into a bacterial cell, then cloned.

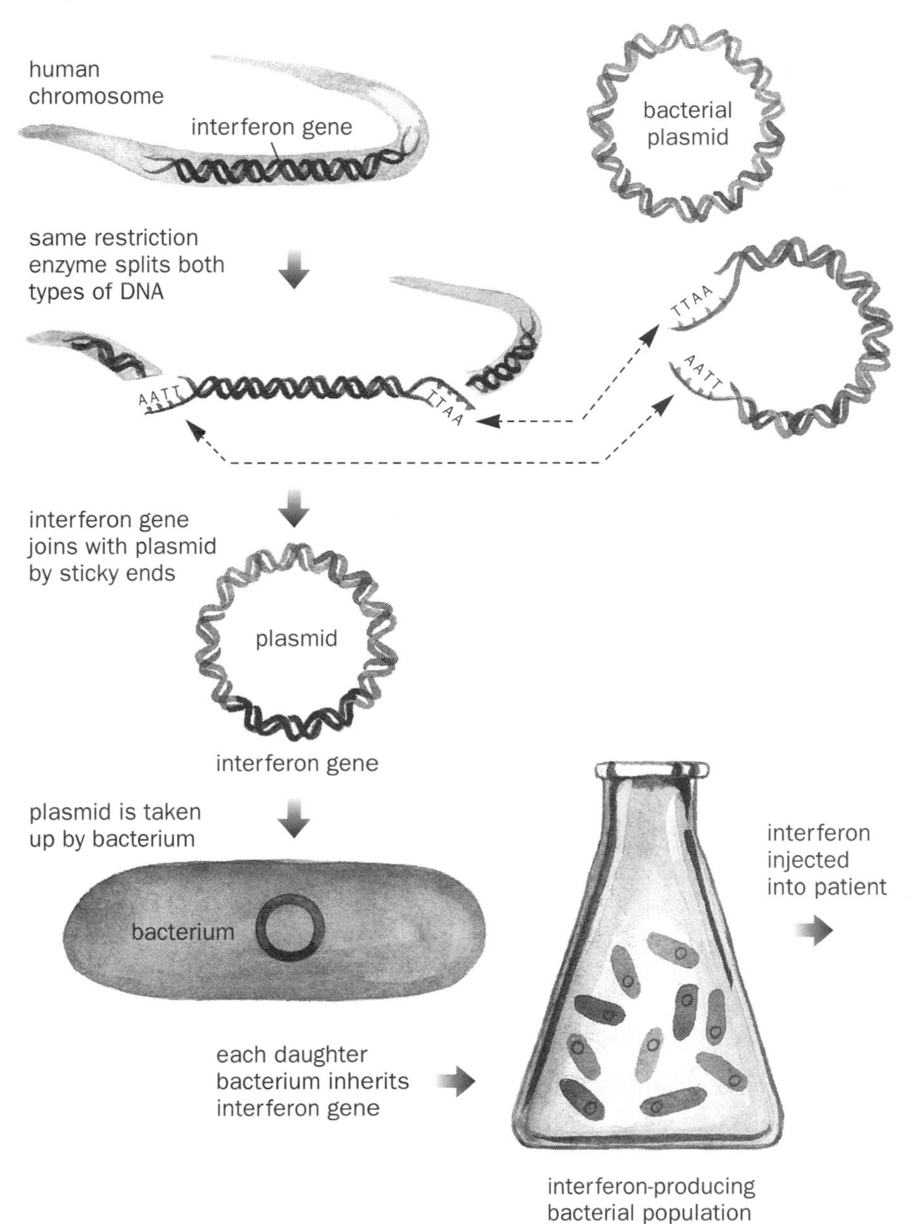

Putting recombinant DNA into bacteria is also, in effect, a simple method of cloning genes. One or two bits of recombinant DNA aren't much use if your aim is to turn out large amounts of the gene product. You need identical copies of the gene in millions of cells. The easiest way to achieve this is to engineer the gene into a few cells and let them multiply, duplicating the new gene along with their own DNA each time they divide in two.

Gene expression

I've talked about genes until now as if the mere presence of a given gene in a cell is enough to make the cell carry out that gene's instructions. If you think about it, that obviously can't be true. Every cell in your body has the same genes, but the cells aren't all alike. Genetic potential isn't the same as genetic fate. Every cell has far more genes than it uses, and only a proportion of the genes are actually "turned on," or expressed, at any one time, making one cell a heart cell and another a brain cell. Understanding the mechanism of gene expression is critical to controlling the outcome of genetic engineering.

In bacteria, certain genes are turned on or off according to the conditions in which the microorganisms are growing. For example, the bacteria *Escherichia coli* can use either of two sugars, lactose or glucose, for energy. They need enzymes to release energy from these sugars, and their enzyme production is encoded in their genes. If the bacteria are grown in an environment with both sugars, they prefer glucose, and express the genes for the enzymes that let them use that sugar. Digestion of lactose requires one extra enzyme, and only when the glucose runs out do the bacteria switch on their genes for producing this additional enzyme. By tying gene expression to environmental cues, the bacteria don't waste energy and materials making products they don't need.

The control of gene expression in multicelled organisms is much more complex and not yet fully understood. The cluster of undifferentiated cells that make up an embryo soon after fertilization must quickly begin expressing different genes to produce different body tissues and organs. The cells in a particular tissue or organ may also switch genes on or off at different times during growth. For example, the cells in testicles or ovaries don't switch on the genes that result in sex hormone production until the organism reaches puberty.

The switch mechanisms that regulate gene expression include groups of genes called regulatory genes. Unlike the genes discussed so far — which we must now call structural genes — regulatory genes do not code for enzymes or other proteins. Their function is to either promote or inhibit the sequence of events by which a structural gene is translated into a product. Promoter regions are located adjacent to structural genes on a strip of DNA. When genetic engineers transplant genes for making products, they must include the switches that control gene expression as well as the genes themselves.

Cloning plants, animals, and cells

Take a cutting from a plant, put it in a pot of soil, and you have cloned an organism. The plant that grows from the cutting will be genetically identical to the one from which you took the cutting. Its development is made possible because each cell at the cut edge has the genetic potential to develop into any type of plant tissue needed to form a whole new plant.

While whole plants have been regenerated from cuttings for centuries, biologists in the late 1950s discovered that whole plants can be regenerated from individual cells (see Figure 2.6). Plant cells seem to retain the potential to express any of their genes and thus repeat the developmental process from a single cell to a whole plant. Biotechnologists have taken great advantage of this characteristic of plant cells.

The number of cells you can take from a plant is obviously much greater than the number of cuttings. This offers the possibility of rapidly developing new strains of

Figure 2.6

Cloning from plant cells

Cloning does not necessarily involve genetic engineering. In plant regeneration from individual cells, young cells from a root tip can each be encouraged to form a new plant. The process shown here can be used along with recombinant DNA techniques to quickly develop new strains with particular properties.

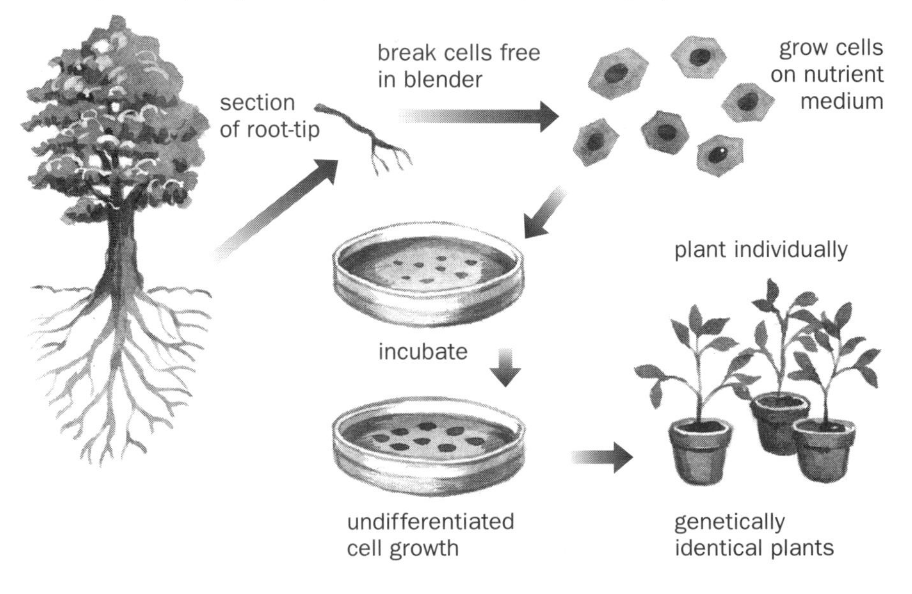

crops or trees from a single plant that has desirable traits.

Among animals, the extraordinary ability of any body cell to give rise to a whole new identical organism has been demonstrated in several species by a technique called nuclear transplanting. The procedure involves destroying the nucleus of an egg cell, then replacing it with the nucleus taken from any cell — say, a skin cell — of another individual. The egg with the transplanted nucleus goes on to develop into a complete new organism identical to the one that supplied the skin cell nucleus. This technique has already been used to clone frogs and mice.

Cell cloning occurs naturally each time a cell divides in two. Most cells divide a certain number of times and then dic. The number of cell generations they produce is genetically determined. Cancer cells, however, are a special case. They seem to have lost their control over duplication and continue to divide and copy themselves indefinitely.

The origin of this condition seems to lie in a mutation that affects the regulatory genes. This mutation makes cancer cells all but immortal, their genes for promoting cell division stuck in the "on" position. Many cancer cells also contain specific tumor-causing genes called oncogenes, which are associated with the uncontrolled cell proliferation that produces tumor growth.

Biotechnologists use the cancer cells' property of unstoppable growth to advantage by joining cancer cells to cells that make desirable products. The hybrid cells that result from this marriage combine the cancer cells' proclivity for endless multiplication with their partner cells' production of enzymes, hormones, or whatever else is chosen. Cultures of such fused cells, called hybridomas, are used to mass-produce huge quantities of valuable proteins.

Monoclonal antibodies

Probably the most important products now derived from hybridoma technology are monoclonal antibodies, whose development won George Koehler and Cesar Milstein a Nobel prize in 1984. Antibodies are proteins produced by certain white blood cells to fight infection. Each antibody is specific to a particular foreign particle invading the body, such as a bacteria or virus. They inactivate the invaders by attaching themselves to them.

Obviously it would be of great value to medicine if antibodies could be produced in the lab in large amounts. That possibility was always limited, however, by the fact that white blood cells do not survive for very long outside the body. To overcome this problem, Koehler and Milstein "persuaded" some white blood cells to fuse with cancer cells taken from tumors. Fusion is not something cells normally do, but it can be promoted by using chemicals, viruses, or, more commonly nowadays, placing the cells together in a high-frequency electrical field. Hybridoma cells obtained in this way multiply and turn out a continuous supply of antibodies — called monoclonal because they are all descended from one original cell.

Although monoclonal antibodies were originally developed for use in medicine, they have subsequently found numerous applications as precise seek-and-find tools. Designed to attach themselves to one type of particle, and one only, monoclonal antibodies have the ability to unerringly locate and mark any given target in any quantity in a mixture as complex as you care to produce. This quality makes them invaluable for a number of uses, such as the analysis of chemical mixtures, monitoring of particular substances, drug manufacturing, and the diagnosis and treatment of certain diseases. For example, monoclonal antibodies targeted to tumor cells can be used to detect the presence of cancer long before symptoms appear. The same antibodies can be "armed" with drugs, which they will carry through the body directly to the cancer cells, selectively destroying them without harming other cells.

DNA probes

Suppose you want to find where the human gene that's responsible for producing insulin is located on the chromosomes. You could break up each chromosome into fragments, combine each fragment with a plasmid, insert the recombinant plasmids into bacterial cells, and check to see which bacteria make insulin. But the chance of any random fragment having the gene you want is remote, and to test all fragments in this way would be time-consuming and costly.

A much quicker way of doing this sort of research is to work backwards from the gene product to the gene. If you know the sequence of amino acids in the protein encoded by the gene, you can use the universal genetic code to translate this into the base sequence needed to string those amino acids together. The base sequence, of course, is your gene.

From there, you turn to another ingenious automated machine that can readily assemble up to two dozen or so nucleotides on a template in any sequence requested. Synthetic single strands of DNA made in this way are used as probes to locate specific genes on chromosomes. For this purpose, all that's needed is a short, distinctive sequence of several nucleotides complementing a sequence on the gene of interest. The probes are made radioactive for later identification, then added to the DNA sample to be analyzed. After the probes have paired with their corresponding genes on the DNA, their location is pinpointed by spreading the DNA carefully over a sheet of special filter paper and placing the filter paper in contact with X-ray film. Each probe and its bonded gene show up as a dark spot.

DNA probes are used for such things as mapping the distribution of genes on chromosomes, locating the presence of recombinant DNA in bacterial cultures, or finding oncogenes on a person's chromosomes, giving advance warning of cancer risk. They are also part of the technology involved in the process of DNA "fingerprinting."

DNA "fingerprinting"

Also known as DNA profiling, this is the technique you most often read about in connection with criminal cases. Developed by Alec Jeffreys in England in the early 1970s, the process is based on the fact that the distance between restriction cleavage sites on DNA strands (that is, the sites at which restriction enzymes make their cuts) differs from person to person. If you cut up DNA samples from any two people using the same restriction enzymes, you end up with two different assortments of DNA fragments of different lengths. Each particular assortment is unique to an individual, like a fingerprint. The occurrence of many patterns of fragment sizes is called Restriction Fragment Length Polymorphism (RFLP).

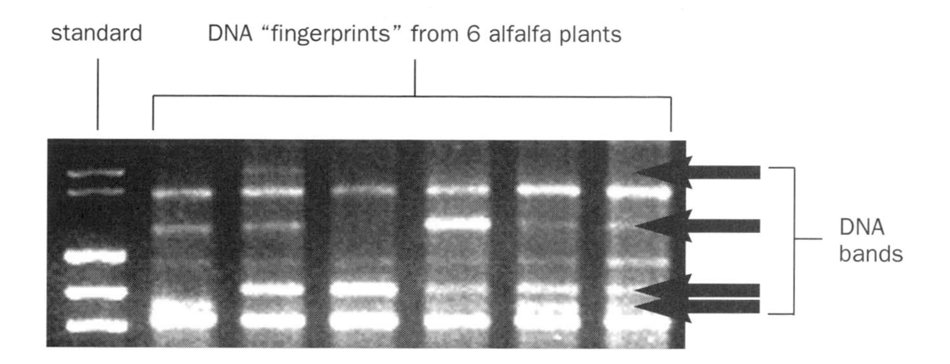

Figure 2.7 Fragments of DNA from a sample line up to produce a distinctive DNA "fingerprint." Whether human being or alfalfa plant, each of us is unique.

To analyze RFLPs, a sample of fragmented DNA is placed in a tiny well in a gel. When exposed to an electric field, the fragments migrate through the gel and are separated according to size. The fragments are then transferred to a nitrocellulose sheet, which holds them in place while a radioactive probe is added. The probe bonds to all fragments containing a specific sequence. X-ray film is placed against the sheet and then developed. A series of bands like a product bar code, on the film (Figure 2.7), reveal the location — thus size — of the labelled fragments. The pattern of bands can be reliably used to identify the individual from whom the DNA sample was obtained.

The exciting part of the DNA "fingerprinting" technique is that it can be used on any substance containing genetic material — blood, saliva, semen, skin, hair, or other tissues. More recent "fingerprinting" techniques in conjunction with PCR technology (the DNA copier) have made it possible to identify individuals from samples as scanty as a few droplets of saliva left on a telephone mouthpiece or a single hair follicle.

Polymerase chain reaction

"How does it work?" I asked the woman, nodding at the small black machine. "It uses a heating-cooling cycle," she began. "It heats up the DNA and melts it so the two strands come apart. Then it cools down and lets each strand build a complementary strand. Then it heats up again and splits them, and so on. It doubles your DNA each time."

Of course, there's more to it than that simple version allows (see Figure 2.8), but there's the gist of it — enough to give an insight into the remarkable extent to which technology has developed in the last few years. The PCR technique was pioneered by Dr. Kary Mullis, who won the Nobel prize for this feat as recently as 1993. But the discovery that was a research breakthrough only a few years ago is already set to become almost commonplace in hospitals and courtrooms.

Polymerases, by the way, are a class of enzymes involved in the processes of building new strands of DNA or RNA. The DNA polymerase used in the PCR machine is derived from bacteria that live in hot water springs. They are among the few enzymes that can function at the high temperatures (close to boiling point) needed to split apart DNA molecules. It is another reminder that biotechnology borrows its tools from nature, although many people see the uses to which those tools are put as decidedly unnatural. The next four chapters look at some of those uses in a number of different fields.

Figure 2.8

Polymerase chain reaction (PCR)

The polymerase chain reaction is used to copy genes. The process works by heating DNA strands to separate them, then adding primers. These are short sequences that attach to the sequence flanking the end of a gene and promote the building of a new gene. The cycle is repeated every few minutes.

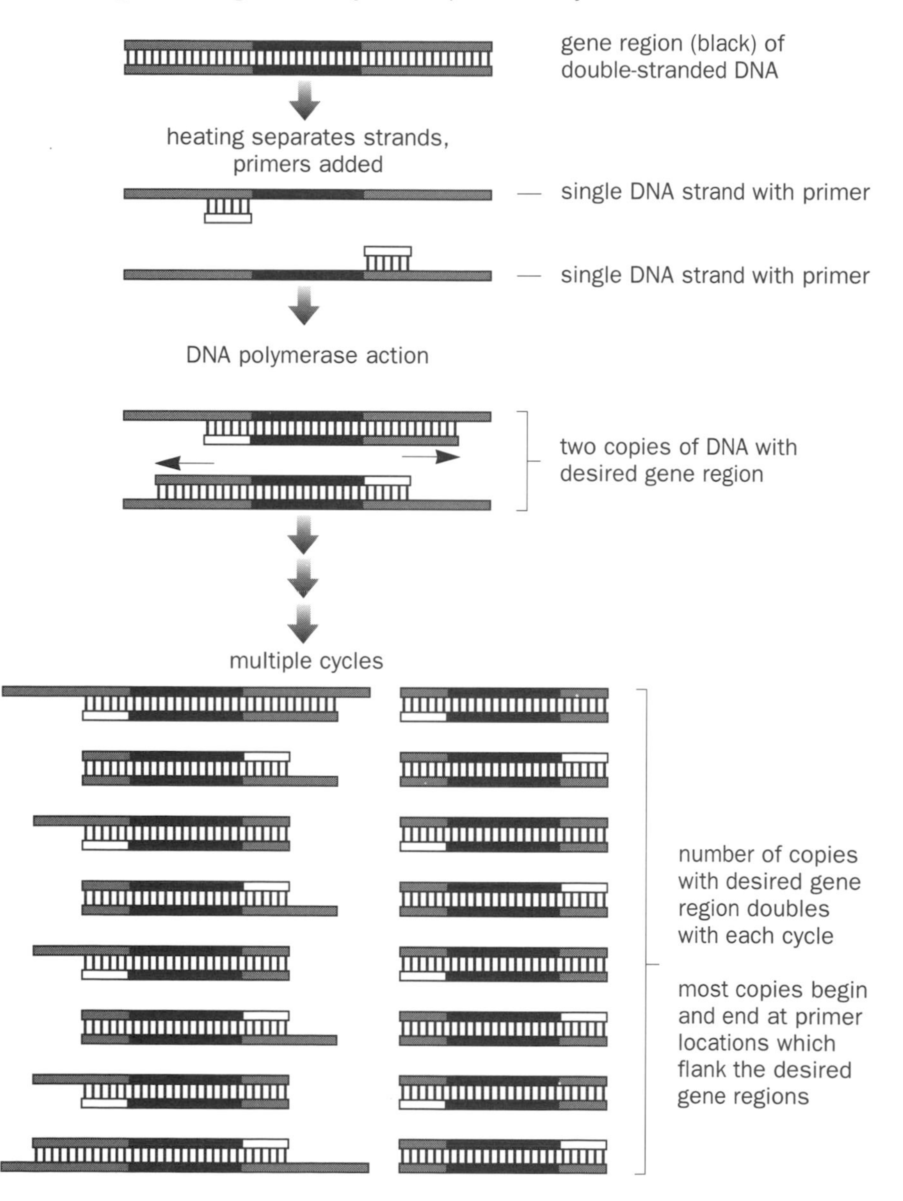

Chapter 3
Biotechnology and the Body

Who would have thought that healing the sick was so fraught with controversy? When less was known about how the body works, and what causes diseases, people had little choice but to accept illness and disability as part of their fate. Depending on a person's philosophy of life, good health was a question of good luck or good morals, disease a misfortune or a punishment for wrongdoing.

Today, we are not inclined to accept fate. Our bodies are less like reflections of our souls, more like consumer products to be properly maintained and repaired, inside and out, by the latest tools on offer in the medical marketplace. While moralizing still attends the sickbed, the questions now deal in economics, rights to privacy, freedom of choice, priorities, and (reflecting the materialism of our culture) ownership of body parts and information. From having little say over our medical well-being, we may now, thanks to biotechnology, have a surfeit of options.

Many of the fears and hopes people have about biotechnology come together in their most potent mix in the field of medicine. Bioengineering techniques give physicians powerful new ways to treat some disorders, but might these same techniques paradoxically devalue human life, reducing us to collections of medical raw materials? Genetic analysis offers sophisticated advances in early detection and diagnosis, but does this only give more ways to tell us our likelihood of getting a particular disorder, without offering any real remedies?

The medical industry is today's biggest customer for biotechnology. The industry includes everything from physicians in hospitals to manufacturers of every kind of equipment, diagnostic techniques, drugs, hormones, vaccines, and other biochemicals. While each addition to the health-care arsenal may be cause for comfort to present and future patients, many applications raise questions that go beyond the scientific and technical, bringing social, economic, ethical, and legal issues in their wake.

New parts for old

If we view the body as an assembly of parts, it doesn't seem odd when we treat faulty or worn-out parts by replacing them. Physicians first achieved limited success doing this as far back as the 1800s, grafting pieces of fresh skin onto burn victims, but it wasn't until well into the 20th century that scientists discovered the secret to transplanting entire organs.

Organ transplants were a psychological as well as a surgical breakthrough, a step towards understanding the body by literally deconstructing it. While a body might be more than the sum of its parts, transplant techniques proved that a single organ could be responsible for disease in the whole. These developments laid the groundwork for the finer probings of biotechnology, which ultimately aim to give us a complete molecular inventory of ourselves.

The first successful organ transplant was carried out in 1951, when a kidney survived the move from one body to another and kept working. In the mid-1990s, kidney transplants have become almost a commonplace triumph of surgery, with about 100,000 defective kidneys being replaced around the world every year. Several former patients have already lived more than 20 years with a kidney they were not born with.

Liver and lung transplants were first attempted in the 1960s, but no operation caught the public's attention more than the first heart transplant, carried out by South African surgeon Dr. Christiaan Barnard on December 3, 1967. Like the first human in space, the pioneering physician made science into headlines and became a household name. Like the space program in the '90s, heart transplants now merit barely a paragraph on an inside page of the newspaper. In the United States today, where most of these operations take place, more than 2,000 hearts each year are set beating inside new chests.

One of the key discoveries that helped make organ transplants possible in the first place was the concept of matching tissues. Unless the tissue of a donor matched that of the recipient very closely, a transplant was likely to fail. The patient's body, able to distinguish between the tissues of "self" and "non-self," would reject the new graft as an unwanted invader.

British researchers Frank Burnet and Peter Medawar first realized the importance of matching tissues during the 1940s after studying how the immune system works. They found that the body responds to invading foreign materials, such as viruses, by producing white blood cells to attack and destroy them (Figure 3.1). The researchers used this observation to explain why early attempts at transplants failed: the patient's body treated new tissues like germs. The more dissimilar the tissues of donor and patient, the greater the immune response. The solution was to suppress the immune system and to use tissues from a blood relative as closely genetically matched to the patient as possible.

The first drugs used to suppress immune reactions in transplant patients were crude. They worked by killing bone marrow tissues where new blood cells are formed. Loss of bone marrow caused terrible side effects and left patients vulnerable to all kinds of infections. As a result, survival rates from early transplant operations were low.

A far superior and more precise tool came to light in 1970, transforming the entire field of organ transplant surgery. The "wonder drug" cyclosporine is produced by a fungus found in soil. It helps prevent organ rejection by inactivating the body's T-cells — one of the types of white blood cells active in the immune system. As well as leaving the rest of the immune system unharmed, cyclosporine has few side effects. Its impact stops when the drug is no longer taken, allowing the patient's body to return to normal. After cyclosporine was first used on humans in 1978, the survival rate of liver and kidney transplant patients doubled, and rejection of heart transplants was practically eliminated.

The success of immune system drugs made transplant operations safer and more effective, but in turn raised a different problem — that of organ supply. With a sharp increase in demand for spare body parts, 10 or more patients were now waiting for every organ made available by a suitable donor.

Figure 3.1

How the immune system works

The immune system has two branches, using two main types of white blood cells. T-cells respond to foreign materials by becoming killer cells and memory cells. B-cells produce antibodies and memory cells. Memory cells help the body respond more quickly to subsequent invasions by the same foreign material.

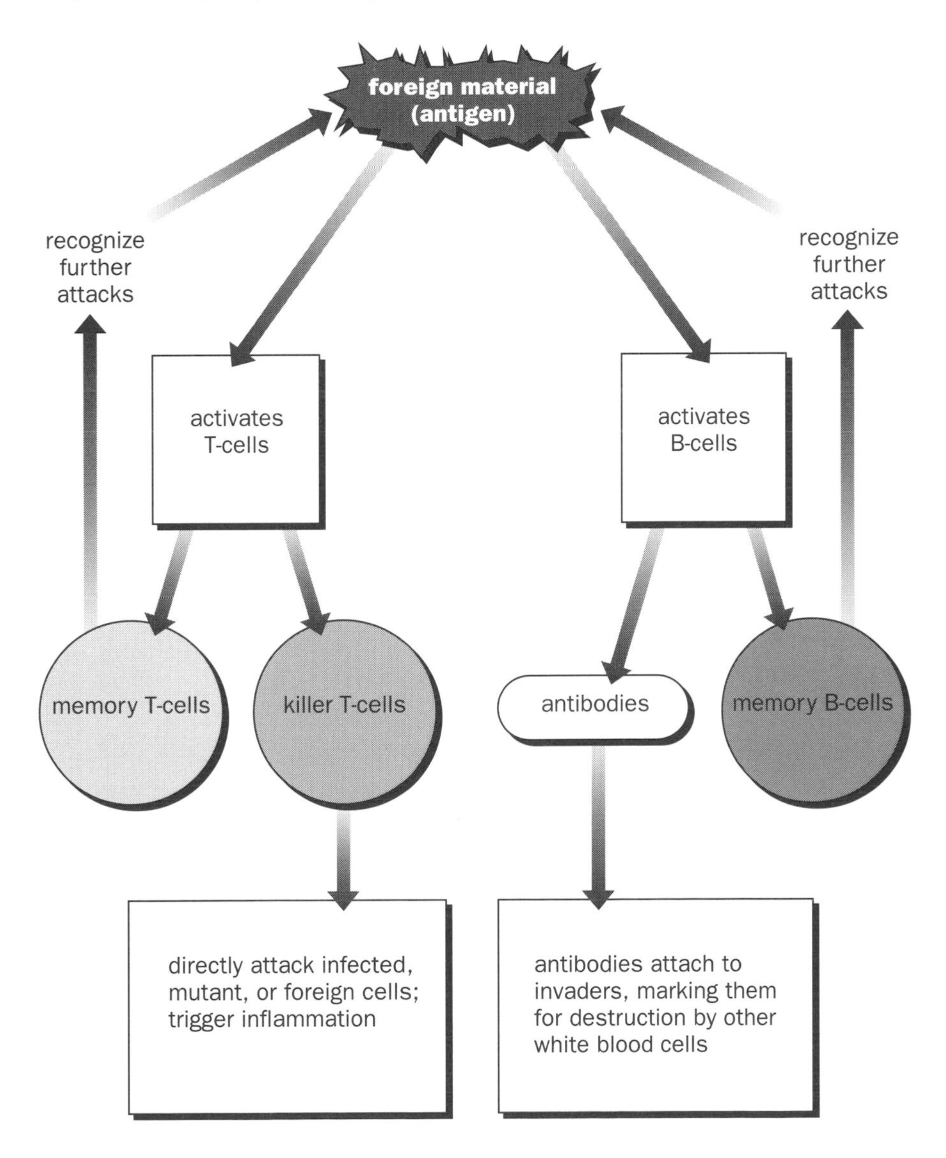

Seeking to meet the demand, scientists explored other sources of supply, including different species. In 1984, for example, the fatally defective heart of a 12-day-old infant known as Baby Fae was replaced with the heart of a young baboon. Despite the use of cyclosporine, the baby rejected the heart and died 20 days after the operation. In 1992, a baboon's liver was used to treat a man dying of hepatitis. Similar operations followed soon after, using organs from baboons and pigs.

The long-term failure of these experiments, as with early transplants, comes back to the question of matching tissues. Now, however, instead of waiting for a close genetic match, scientists aim to produce one by genetic manipulation. In 1994, pigs were engineered with human genes so that their tissues produced human proteins that inhibit organ rejection. Hearts from the genetically altered pigs were subsequently transplanted into baboons (as models for humans). The primates survived only 19 hours after the operation, but they did much better than a control group of animals given regular pig hearts, which survived no longer than 40 minutes.

A different option for replacing some body parts is to use totally artificial structures, bypassing the need for donors of any kind. Thus, the heart's function of moving blood around the body can be carried out by a plastic and metal pump — surgery's greatest vindication of the materialist view of the body. This kind of development shifts attention from transplants to implants: fabricated body parts that can be manufactured. Included among implantable parts are such items as pacemakers, stainless steel hips, artificial lenses, and various other prosthetics. Such implant operations outnumber transplants in the United States by 100 to 1.

Although in most cases the synthetic materials used to make implants do not produce an immune reaction, implants have resulted in auto-immune disorders in one well-known instance. Of the more than one million women who have received silicone-gel breast implants since the early 1960s, several thousands subsequently developed health problems related to the operation.

Perhaps the most intriguing possibilities for replacing body parts lie in attempts to manufacture new biomaterials and grow new organs from scratch. The idea is to take individual cells from a tissue or organ and seed them in a fine mesh of soluble material, then incubate them until they multiply and connect up. The resulting tissue is implanted in the body where it establishes connections with the patient's tissue, taking hold while the mesh dissolves away. If the patients' own cells are used to start new tissue, there won't be any immune reaction, or any need for drugs to control rejection. In an account of this research reported in the July 8, 1995 issue of *Science News*, Gail K. Naughton, chief scientist for Advanced Tissue Sciences in La Jolla, California, predicts a big future for the technology. She claims that "tissue engineering today is where genetic engineering was 10 years ago."

So far, tissues such as skin, cartilage, and tendon have been grown in this way outside the body and then transplanted into animals. Parts of organs such as the liver, kidney, and pancreas have also been grown. In 1994, artificial livers, made outside the body from cloned liver cells wrapped in synthetic material, kept 9 of 12 patients alive at King's College Hospital in London until donor livers became available for transplant.

Fifty years ago, the development of transplant techniques was pioneering work. Today, transplant surgery is well established, and a new generation of physicians are setting their sights on another frontier. For these medical researchers, the focus is not organs and tissues but molecules. For them, diseases and disorders are only the final expression of chemical interactions inside cells. The way to control illness, goes their argument, is to control the commanders of the cells — the genes.

It's all in the genes

A length of DNA — that's all it takes to make one individual different from another. One person may be born to live a robust life into healthy old age, while another suffers a crippling disability leading to an early grave. According to this view of things, our state of health is already circumscribed by genetic lottery months before we even began to draw breath, primed by an inheritance that will eventually give us one disease or another. Biotechnology's big promise is to subvert genetic destiny and cure the previously incurable; to rig the lottery and make everyone a winner.

Unless your family name is Alzheimer or Tay-Sachs, you may think inherited disease is other people's problem. But scientists are turning up genetic links to diseases and other medical conditions all the time. Over 4,000 medical disorders are already known or suspected to be caused by defective genes, many of them maladies that appear only later in life. Some are caused by single genes, others develop from the interaction of two or more genes, and still more are due to a combination of genetics and environmental agents such as foods or toxic chemicals. Genetic flaws produce nearly all miscarriages, one-fifth of all infant deaths, and 80 percent of all mental retardation.

While some distinguished scientists go so far as to claim that all human disease is ultimately genetic, not everyone agrees with such an extreme interpretation. Environment, diet, and habits can all modify the outcome of many genetic predispositions. But neither can we assume any longer that genetic illness isn't our concern, that we don't harbor in our bodies a mutant molecule that may at some point afflict our lives, or the lives of our children.

Take cystic fibrosis as an example. One of the most common genetic disorders, it is produced by a gene carried by one in every 25 people of North European stock. (Having a copy of the gene isn't the same as having the disease. It takes two copies — one from each parent — for the disease to be manifested.) The mutant gene affects the ability of cells to secrete normal products. Like a vital cog missing from an engine, this one small defect makes all the difference between smooth functioning and calamity. The mutant gene causes development of cysts and fibrous tissues in the pancreas; degeneration of sperm-producing cells, causing sterility; and production of thick, sticky mucus in the lungs, which often proves fatal.

Examples of conditions known to be gene-based

Alzheimer's disease
amyotrophic lateral sclerosis (Lou Gehrig's disease)
arthritis
asthma
cancers
cystic fibrosis
diabetes
Down syndrome
hemophilia
high blood pressure
hypercholesterolemia
multiple sclerosis
muscular dystrophy
neurofibromatosis
schizophrenia
sickle-cell anemia
spina bifida
Tay-Sachs disease

Scientists now know from DNA probes exactly where the gene responsible for cystic fibrosis is located on the chromosomes. They can identify carriers of the gene and counsel them about their chances of passing on this disease to their children. They can also identify the gene in embryos, allowing parents to choose whether to continue a pregnancy if the embryo proves destined to suffer from the disease. Another option recently opened up by biotechnology is gene therapy, which aims to repair genetic damage by giving patients healthy copies of the problem gene.

The beginnings of gene therapy

The story of the race to carry out gene therapy is engagingly told in the book *Altered Fates*, written by two prizewinning journalists, Jeff Lyon and Peter Gorner. Like James Watson's 1968 book, *The Double Helix*, the account makes interesting reading not only for its descriptions of important scientific discoveries, but also for its portaits of the personalities involved, and its revelations about the behind-the-scenes politics of high-profile scientific research.

The steps leading to the first legally approved operation in which a human patient was given engineered genes from another species are worth summarizing, as an example of how things can move from the theories of research scientists to the end of a hollow needle in an operating room.

In 1986, medical researchers discovered that some peculiar white blood cells had invaded and killed early tumor cells in cancer patients. They named the blood cells tumor-infiltrating lymphocytes (TILs). Unfortunately, as the tumors subsequently spread, the TILs became overwhelmed and lost the fight. To boost the body's own defense, doctors took TILs from patients, cultured them in the lab, then returned them in huge numbers to swamp the patients' lymph and blood system. About half the people with terminal melanoma (skin cancer) treated this way responded well.

The difficulty researchers had at this point was to know why some patients lost their tumors after treatment with TILs while others didn't respond at all. To find an answer they had to track TILs inside the body. However, no monitoring methods available at the time could mark the cells for longer than two weeks. After that, their fate was a mystery.

The solution eventually came as the result of a one-hour meeting between the leading TIL researcher, Steven Rosenberg (incidentally, the physician who diagnosed Ronald Reagan's cancer in 1985), and a pioneer in gene transfer experiments, William French Anderson. The two men, unaware until then of the details of one another's work, quickly devised a scheme to use altered genes for tracking TILs.

The plan was to splice a bacterial gene for resistance to a particular antibiotic into a virus, then culture the engineered virus with TILs. The virus would infect the TILs and transfer its genes onto their chromosomes. As the TILs later multiplied in culture, each would carry the telltale gene of antibiotic resistance. With this added marker, the TILs and their descendants could be tracked in tissue samples anywhere, anytime — they would be the only cells that survived when soaked in the antibiotic.

Figure 3.2

Keeping track of engineered genes

Finding engineered cells among a culture of hundreds of thousands of regular cells is like looking for a needle in a haystack. A common technique is to attach a genetic marker to the recombinant DNA. A bacterial gene for antibiotic resistance was used as a marker to track engineered tumor-infiltrating lymphocytes (TILs) in the first experiment in gene transfer to humans.

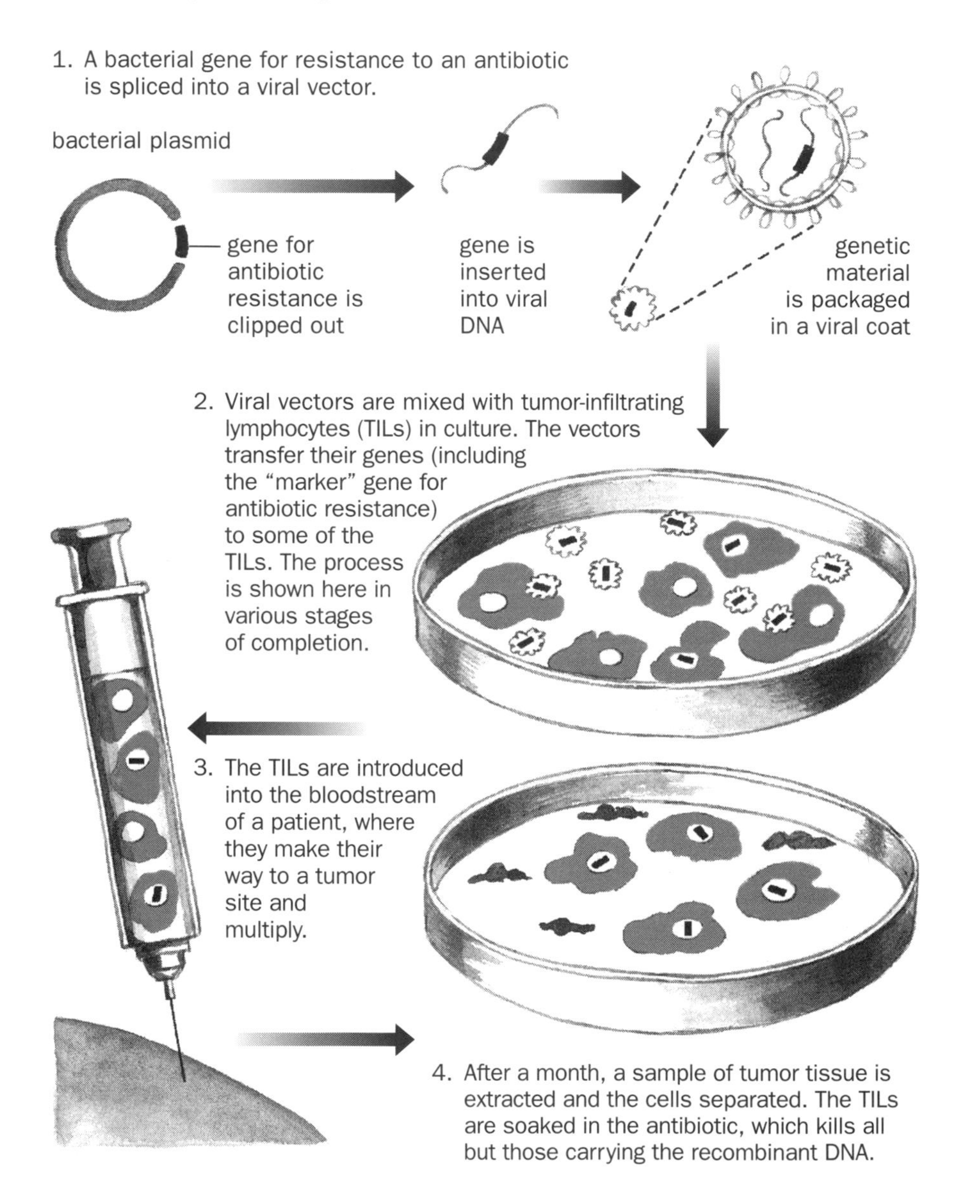

Retroviruses

The viruses used as vectors in gene therapy are a special class called retroviruses. They are simple viruses whose genetic code is written in RNA rather than DNA. Working in reverse order to the usual DNA-to-RNA-to-protein sequence, they first use their RNA as a template to build a complementary single strand of DNA. This process is initiated by an enzyme called reverse transcriptase. A retrovirus may have only three genes coding for proteins: one for its envelope proteins, one for its core proteins, and one for the reverse transcriptase enzyme.

When using retroviruses as recombinant vectors, scientists remove the viral genes and replace them with therapeutic genes. The altered virus can still transfer the added gene to the host chromosomes, but is no longer capable of replicating itself.

Elegant and simple, the technique (routinely used by bacteriologists and shown in Figure 3.2, opposite) would tell surgeons how well TILs survive to keep on attacking tumors. And it would tell genetic engineers whether altered genes could continue functioning and reproducing inside a human body. The introduced genes would not be doing anything actively therapeutic (the eventual goal of gene therapy); they would serve only as marker tags on the tumor-killing cells. But it was a start. It would show for the first time whether you could implant foreign genes into a human without any of the disastrous consequences that some people predicted.

The next step was to get the experimental treatment reviewed by the National Institutes of Health (NIH) and the Food and Drug Administration (FDA). This initially involved a series of meetings with different subcommittees set up to approve protocols for human gene therapy. Stormy debates at these hearings separated molecular biologists on one side of the arguments from medical researchers on the other. Differing fundamentally in their perspective, and in their assessment of risks and benefits, the pure scientists were more wary than the physicians, who worked with dying and desperate patients. Cautious attitudes won the early skirmishes, but lost the war at a decisive and historical meeting of the full Recombinant DNA Advisory Committee (RAC) held on October 3, 1988.

In the end, emotion rather than strict science carried the day. While several members of the RAC were still concerned about the lack of data from similar experiments on lab animals (mouse TILs would not accept genes from viral vectors), the committee was eventually persuaded on balance by the thought of cancer patients whose time was running out. As Harvard Medical School microbiologist Bernard Davies put it: "It is virtually not possible to have more risk than certain death." The committee voted 16 to 5 to approve the experimental treatment, the dissenting voices coming from molecular biologists.

The patient finally selected for receiving gene-marked TILs was 52-year-old Maurice Kuntz, diagnosed with malignant melanoma, a deadly form of skin cancer that had spread to his liver. His prognosis: two months to live.

The cancer of sun-lovers

About 32,000 Americans develop skin cancer every year and nearly 7,000 die from it. In Australia, where sun worshipping abounds, skin cancer mortality rates more than quadrupled between 1931 and 1977, but now seem to be stabilizing.

In some ways, the rest of the story is an anticlimax. There was no miraculous recovery, although Maurice Kuntz did live for nearly a year after initial treatment in May 1989 — far longer than expected. For the researchers, results were mixed. Genetically altered TILs were detectable in his body up to three months after injection, giving more information about their behavior. More importantly, the operation broke through the psychological barrier that until then had kept people from tampering with human genes. The deliberate insertion of a gene into a human body came about 20 years after scientists first started to cut and recombine genes in fruit flies, mice, bacteria, toads, tomatoes, and other species. Humans aren't basically different from other forms of life after all — at least as far as their DNA is concerned.

The Human Genome Project

The Human Genome Project is an ambitious plan to map and sequence all 100,000 or so genes found in human DNA. It is a task that has occupied hundreds of scientists in labs around the world since about 1986. The first human genes to be identified, back in the 1970s, were those connected with diseases such as cystic fibrosis. Part of the motivation to sequence the entire genome (that is, all the genes present in a complete set of chromosomes) was the desire to learn more about the genetic roots of disease and to discover more genes that might be used in gene therapy. In 1971, only 15 human genes had been localized to specific chromosomes, most of them on the easily identified sex chromosome. By the mid-1990s, researchers had mapped the location of about 2,000 genes: an impressive feat, but still only two percent of the entire human genome.

The ability to map genes was boosted by the development of recombinant DNA technology — in particular the use of restriction enzymes to cut DNA molecules into small fragments with known endpoints. The restriction enzyme cutting sites act as easily identified markers that let scientists compare different fragments for the presence or absence of particular genes. Bit by bit, they build up collections of fragments that overlap each other in known order until they have eventually spanned the entire length of a chromosome. Adjacent fragments form ordered chromosome libraries that help researchers locate particular genes.

Mapping the location of genes on a chromosome is, however, only the first step. The ultimate aim is to know the sequence of bases in each gene. This is an even lengthier task, since there are about three billion base pairs in a set of 23 human chromosomes.

Many scientists found the launch of this huge research program stimulating — like the American drive to put a man on the moon during the 1960s. Reviewing the genome project in 1989, James Trefil, professor of physics at George Mason University, wrote: "It represents nothing less than the ultimate scientific response to the Socratic dictum 'Know Thyself'." Other scientists were less enthusiastic, seeing much of the exercise as a colossal waste of time, money, and human resources.

Critics of the genome project argued that the complete sequencing of each gene is simply unnecessary and tedious — "like mapping every tree in Borneo." For medical purposes, the simpler identification of genes responsible for disease is all that is needed. Furthermore, most of the genome is not, in fact, made up of genes that encode protein production. Long stretches of DNA have other functions, such as turning genes on and off, or helping cells duplicate genes during division. Other regions may simply be evolutionary baggage with no useful function, like the human appendix. Professional scientists showed their human side over the debate in passionate letters to learned journals and shouting matches at conferences.

Another concern, shared by people at large, was that full knowledge of the human genome was a scary sort of power, evoking the story of Frankenstein. Are such worries inflated? Putting things in a different perspective, Joseph Gall wrote in an issue of *American Scientist* in 1988: "[Genetic maps] will be like having a whole history of the world written in a language you can't read." What Gall was pointing out is that mapping and even sequencing genes is only a beginning. That knowledge alone won't tell us the genes' functions. Of the 2,000 or so genes whose locations are mapped today, we know the functions of only a few hundred. And knowing the functions won't tell us how those functions are actually carried out — how genes are expressed and what the biochemical steps are between the coding for a protein and the symptoms of a disease. Although advancing knowledge is rapidly closing in on these areas, we needn't worry just yet about having all the secrets of life.

The continuing story of gene therapy

The first actual use of gene therapy began in September 1990, with the treatment of a child suffering from a rare genetic immunodeficiency disease caused by the lack of the enzyme adenosine deaminase (ADA). ADA-deficient people have persistent infections and high risk of early cancer, and many die in their first months of life. The much-publicized bubble boy, David, had this disease. David lived for nine years in a plastic chamber to prevent contact with viruses, which his immune system could not combat.

As with many other genetic disorders, the root of ADA deficiency lies in the body's inability to produce a key chemical because of a defective gene coding. The disease occurs only in children who inherit defective copies of the ADA gene from both parents. A child who inherits a defective copy from one parent and a normal copy from the other will not have the disease, but may pass on the defective copy of the gene to the next generation.

Researchers had identified the normal ADA gene in human white blood cells cells during the early 1980s. They wanted to see what happened when they introduced copies of this normal gene into a lab culture of T-cells taken from ADA-deficient patients. T-cells were used because they are easy to obtain and grow in the lab, and easy to alter genetically.

After ADA genes were transferred into the T-cells by genetically engineered viral vectors, the cells began to produce the ADA enzyme as predicted. The amount of enzyme produced was about 25 percent of normal, but more than enough to correct the conditions caused by ADA-deficiency. As well, the genetically altered cells had the same lifespan as normal T-cells — longer than the lifespan of uncorrected T-cells from ADA patients. The beauty of this technique is that the desired gene not only remains in the cell as long as it survives, but is duplicated and passed on to all the cell's descendants whenever the cells divide.

The importance of stem cells

The ideal cells for making copies of introduced genes and spreading them quickly through a patient's bloodstream are the stem cells located in bone marrow. They are rapidly dividing cells that produce all the different types of red and white blood cells found in the body, including those that make up the immune system. Because their function is to generate new cells, genetically altered stem cells can be a source of healthy blood cells for the rest of the patient's lifetime. Unfortunately, it is very difficult to isolate stem cells from bone marrow tissue, and attempts to engineer stem cells have not so far resulted in large numbers of genetically altered cells appearing in the bloodstream.

Part of the difficulty is getting access to cells located deep in bone marrow. Some stem cells, however, enter the bloodstream and make up a tiny percentage (one to three percent) of blood cells circulating throughout the body. Researchers are trying to find ways to isolate and concentrate these. If they are successful, engineered stem cells may eventually provide a way of permanently curing most, if not all, genetically determined diseases of the blood and circulatory system.

With the success of the lab experiment, researchers were ready to try out the technique on patients suffering from ADA-deficiency. The first to be treated was a four-year-old girl, then, a year later, a nine-year-old. Early results were encouraging for both children. On a regimen of infusion with ADA gene-corrected cells every one or two months, both patients showed normal levels of active T-cells in their blood after a year, and both developed improved immune function.

Because the altered T-cells won't last forever, it wasn't a permanent cure — that would require using bone marrow stem cells. It was a temporary therapy that depends on regular infusion of engineered T-cells. But, like the use of insulin by diabetics, it allowed these patients to lead relatively normal lives. Within a year of her initial treatment, the first little girl was able to attend school, swim, dance, and ice-skate with her family and friends, with no more risk of catching infections than they had.

Blood cells can be genetically altered and reintroduced to the body by a simple injection into a blood vessel. But what if you want to alter genes in the cells of an organ, such as the liver? One approach is to remove a piece of the liver, divide it into individual cells, insert the appropriate genes into each cell, and transplant the engineered cells back into the patient.

A second approach, according to William French Anderson and others, is to develop smart vectors — ones that can find their own way to diseased tissue inside the body. Rather than inserting genes into cells in petri dishes, the new generation of vectors will be injected directly into patients to carry genes to their targets like guided missiles. This could be achieved by attaching molecules to the vector that recognize specific proteins found on the surface of cells in the target organ.

The type of gene therapy I've described adds normal genes to a patient to produce something the patient lacks due to genetic defects. Another type of gene therapy works in a different way, by obstructing genes that cause disease. In this strategy, called antisense therapy, scientists add a gene that mirrors the target gene — say, one that causes arthritis. The engineered gene produces RNA that complements the RNA of the troublesome gene, binding onto it and blocking its action. So, for example, if the disease-causing gene produces an unwanted protein, antisense therapy will prevent the protein from being formed. If the gene suppresses the formation of a wanted protein, the therapy will allow for normal protein production.

The first stage of gene therapy — identifying genes associated with disease — is fairly well established, thanks to the Human Genome Project. News reports frequently announce the discovery of genes responsible for this or that condition, from Alzheimer's to baldness. Research efforts now concentrate on the second and third stages of the process: delivering genes safely to their targets in the body, and controlling gene expression in the altered cells. These steps are crucial to gene therapy's success, and are likely to take the next 10 to 20 years to develop.

Gene therapy developed from the view that disease is a property of genetic structure and regulation. In this perspective, most therapeutic drugs, in effect, act indirectly on some form of gene expression. If ill health is an outcome of faulty cell production — too little or too much of the right proteins — good health is a matter of adjusting the cells' chemical balance.

Much of modern medical treatment depends on the use of chemicals, and a large part of the medical biotech industry involves producing large quantities of pure drugs tailored for specific tasks. Some are extracted from natural sources, some are manufactured synthetic compounds, but more and more are produced by engineering cells with recombinant DNA.

Microbes in medicine

The pharmaceutical business was using the products of cells long before genetic engineering developed in the 1970s. Interest in the potential use of microbes in medicine was stimulated in 1928 by the discovery of penicillin — the first of four major classes of antibiotics now in common use (the others being tetracyclines, cephalosporins, and erythromycins). The original fuzzy mold that settled on some untidy dishes in Alexander

Fleming's London lab, however, was very different from the organisms that produce the drug doctors prescribe today.

That original wild strain of *Penicillium notatum* yielded only minute amounts of the bacteria-killing chemical called penicillin. Not content to leave nature in its natural state, scientists set about altering the genetic character of the mold using the same techniques of selective breeding used for thousands of years on domesticated animals and plants. Like cows and corn before them, generations of microbes were hand-picked for their qualities and fashioned to perform a particular task.

By 1951, after searching for new strains from different locations and comparing one strain with another, scientists had developed a *Penicillium* capable of making up to 60 mg/L of penicillin (0.008 oz/gallon) — much more than the wild strain. Further years of painstaking work brought even greater improvements in both the quality and amount of antibiotic produced, increasing output by over 300 times to 20 g/L of recoverable penicillin (2.8 oz/gallon).

The variations seen in different populations of mold (or any other species) are due to differences in their genes. Many of these genetic differences arise partly as a result of naturally occurring mutations. To speed up the rate of mutation and increase the variety from which to select improvements, scientists subject microbes to radiation or chemical treatment. Many mutations will be lethal, and some have no clear advantage or disadvantage. A few will result in the kind of change scientists want to see, and these are isolated to form an improved stock used for further breeding.

The success of penicillin was followed by the discovery of many other germ-killing chemicals produced by microorganisms. Today there are more than 100 different antibiotics in use, and more than 5,000 additional compounds made by microbes have been shown to kill or disable harmful microbes.

Although antibiotics lack the dramatic appeal and high-tech wizardry of transplant surgery or gene therapy, they have in their more humble way changed the face of medical science, lifting the scourge of many infectious diseases and saving many millions of lives — which is, after all, the purpose of medicine.

Medicines from plants

Another fruitful source of natural medicines is plants. Many have already proved their value and there is a huge untapped potential for further discoveries, since more than 90 percent of the world's half million plant species have never been tested for their pharmaceutical value. Only 120 prescription drugs worldwide are based on extracted plant products.

Taxol, an anticancer drug made from the bark of the Pacific yew tree, is a recent and much-publicized example of a valuable plant-derived drug. Taxol is currently undergoing clinical trials for treating a variety of cancers, but the major difficulty for researchers has been to obtain the drug in sufficient quantities. Pacific yews are slow growing and rare, and the drug-extraction process is time consuming. The taxol molecule has been made synthetically, but the process is not commercially viable. An alternative method being developed is to chemically convert a similar molecule produced by common yew trees.

A more efficient way to produce natural plant drugs in large quantities would be by using recombinant DNA techniques, but far less research has been carried out on the genetics and biochemistry of plants than of microbes or animal cells. To date, microorganisms are still the main tools bioengineers use to turn out pharmaceutical products.

The interferon story

The first big success story in the commercial production of drugs by genetic engineering was interferon, another naturally occurring compound connected with the immune system. Discovered in 1957, interferon is produced by cells in the human body in response to viral attack. It promotes production of a protein that stimulates the immune system, interfering with the spread of infection.

Although the usefulness of interferon was recognized at once, it could not be marketed for widespread medical use. The chemical is produced by the body in such tiny amounts that it would take the blood from 90,000 donors to provide only one gram (0.03 oz) of interferon, and even then the product would be only about one percent pure. In 1978, a single dose of impure interferon cost about $50,000 to obtain.

All that changed dramatically with the birth of genetic engineering. In 1980, Swiss researchers introduced a gene for human interferon into bacteria, the first time such a procedure had been done with human genes. Cloning millions of bacterial cells from the original engineered one, they were then able to produce a cheap and abundant supply of the previously rare protein. By the mid-1980s, supplies had shot up and pure interferon was being produced for about one dollar per dose.

It was an example of the kind of achievement that makes supporters of medical biotechnology so enthusiastic. Interferon is now used not only to combat viral infection in transplant patients but also to fight other viral diseases (including the common cold), and as an anticancer drug.

Genes and vaccines

A big advantage of using genetic engineering to produce drugs is that it's possible to mass-produce chemicals that might otherwise be difficult and costly to extract, or simply unavailable by conventional means. Another important advantage is that drugs produced in this way are pure and, if made using human genes, fully compatible with use in people. For example, before engineered bacteria were cloned to manufacture human insulin, the main source of this hormone (used to treat diabetes) was the pancreas of cattle or pigs. Although similar to human insulin, animal insulin is not identical and causes allergic reactions in some patients. The human protein produced by bacteria with recombinant DNA, however, has no such effect.

To take another example, vaccines against disease are traditionally prepared from killed or "disarmed" pathogens (disease-causing microbes). They are effective in the vast majority of people, but a small percentage of the population have allergic reactions to vaccines. There is also a very small risk of vaccine organisms reactivating to their former pathogenic state. Genetically engineered vaccines are safer because they contain no living organisms, only the proteins that stimulate the body to develop immunity (Figure 3.3).

Figure 3.3

Engineered vaccine

A safe vaccine against viral disease can be produced by engineering the gene for the viruses' protein coat into bacteria. The bacteria manufacture the viral coat protein, which is then injected to stimulate the body to make antibodies against the virus.

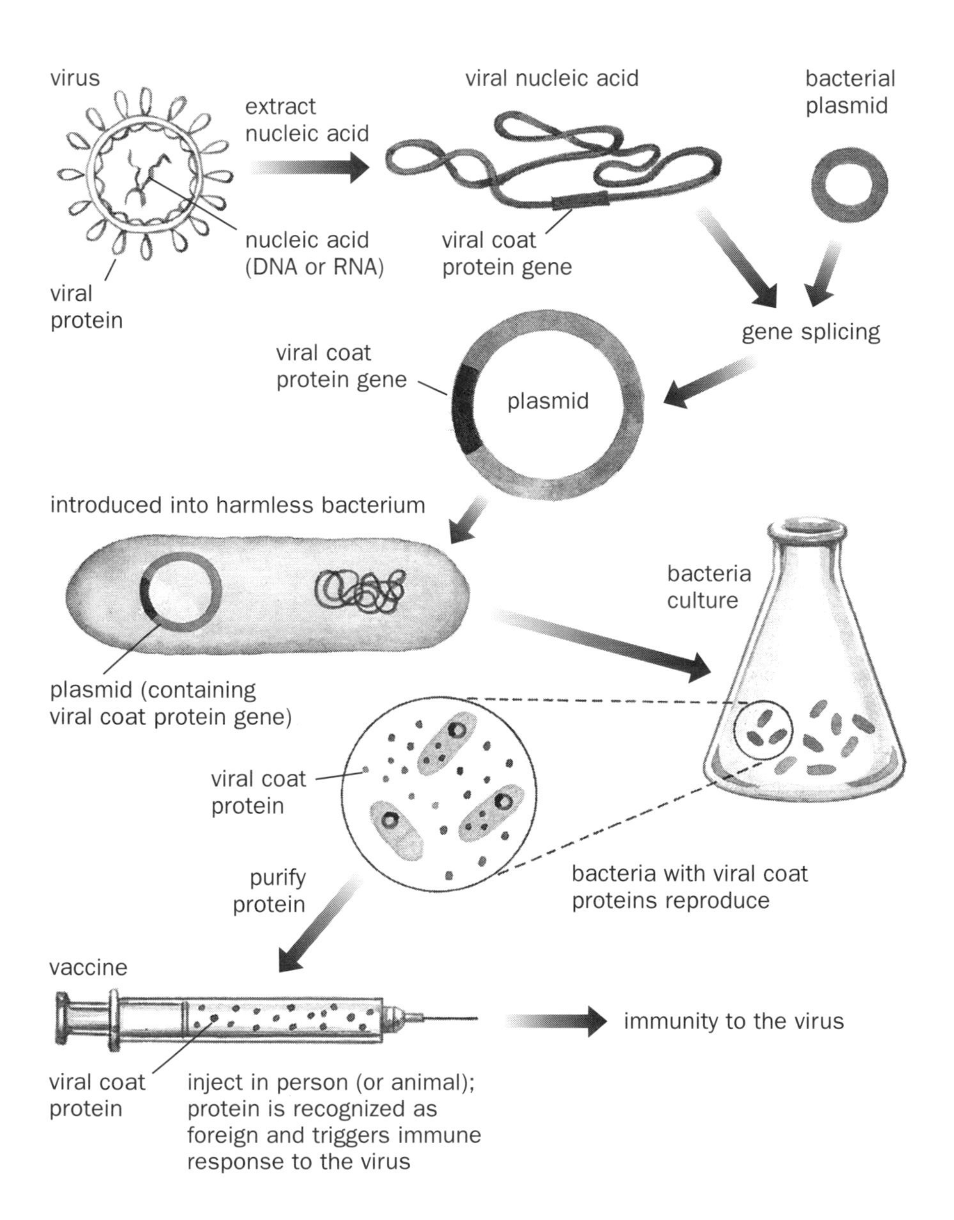

Vaccines are the second-largest category of over 200 drugs now being produced by American pharmaceutical companies using biotechnology. Other products include hormones, interferons, blood clotting factors, antisense molecules, and enzymes. Most of these drugs are still undergoing clinical testing, and are designed to combat cancer, AIDS, asthma, diabetes, heart disease, Lyme disease, multiple sclerosis, rheumatoid arthritis, and viral infections. Outnumbering every other type of biotech medical product, however, are monoclonal antibodies — a versatile group of molecules with seemingly endless applications.

Nature's magic bullets

The ideal drug, said German medical pioneer Paul Ehrlich (1854-1915), would target and eliminate infections while having no ill effects in patients. It would be like a magic bullet, unfailingly hitting the bull's-eye. While no such drugs have yet been made (even the best have occasional harmful side effects), nature makes something very like them in antibodies.

Antibodies are part of the body's immune system. Manufactured by B-cells in the spleen, blood, and lymph glands, antibodies are proteins that latch onto invading microbes or other foreign materials, tagging them for destruction by other body cells. Anything that stimulates antibody production is called an antigen.

Each B-cell produces an antibody molecule shaped to fit precisely to the surface of a particular antigen, like a hand in a glove. Since antigens come in all shapes and sizes, the body is able to produce literally millions of different types of antibodies, each specific to an antigen and

Figure 3.4

Antibody-antigen fit

Antibodies have specific binding sites, which latch onto corresponding sites on the surface of an invading cell.

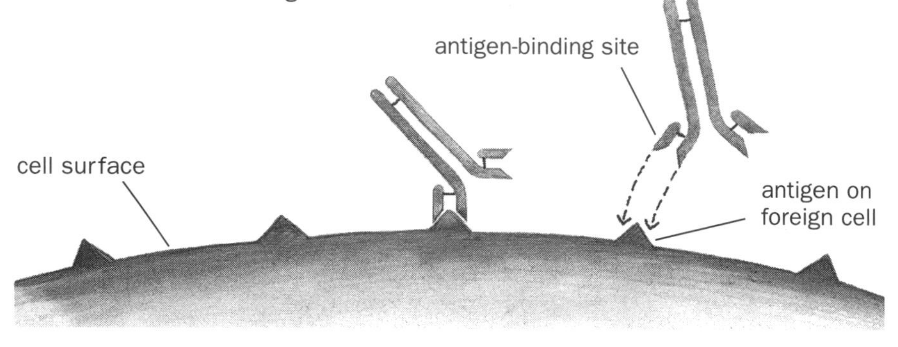

different from the others. A large antigen can stimulate production of a range of antibodies, which attach themselves to different parts of its surface.

Once produced, antibodies continue circulating in the blood in small quantities. This is the basis of the immune response, which prepares the body for further invasions by the same germ. It also allows doctors to discover a patient's past history of infection, by examining the antibodies in a sample of the patient's blood.

The antibody's "seek and find" ability has a lot of potential medical uses, not only to combat diseases but also to help diagnose illnesses and detect the presence of drugs and abnormal substances in the blood. The body produces only minute amounts of each antibody, however, and blood samples have mixtures of many different types. Researchers needed to duplicate large numbers of identical antibodies. The means to do this came with the development of hybridoma technology for producing monoclonal antibodies, described in Chapter 2. Monoclonal antibodies (mAbs) are the products of cloned cell cultures derived from single original B-cells, as shown in Figure 3.5.

Figure 3.5

Making monoclonal antibodies

Monoclonal antibodies are made by fusing antibody-forming cells with tumor cells.

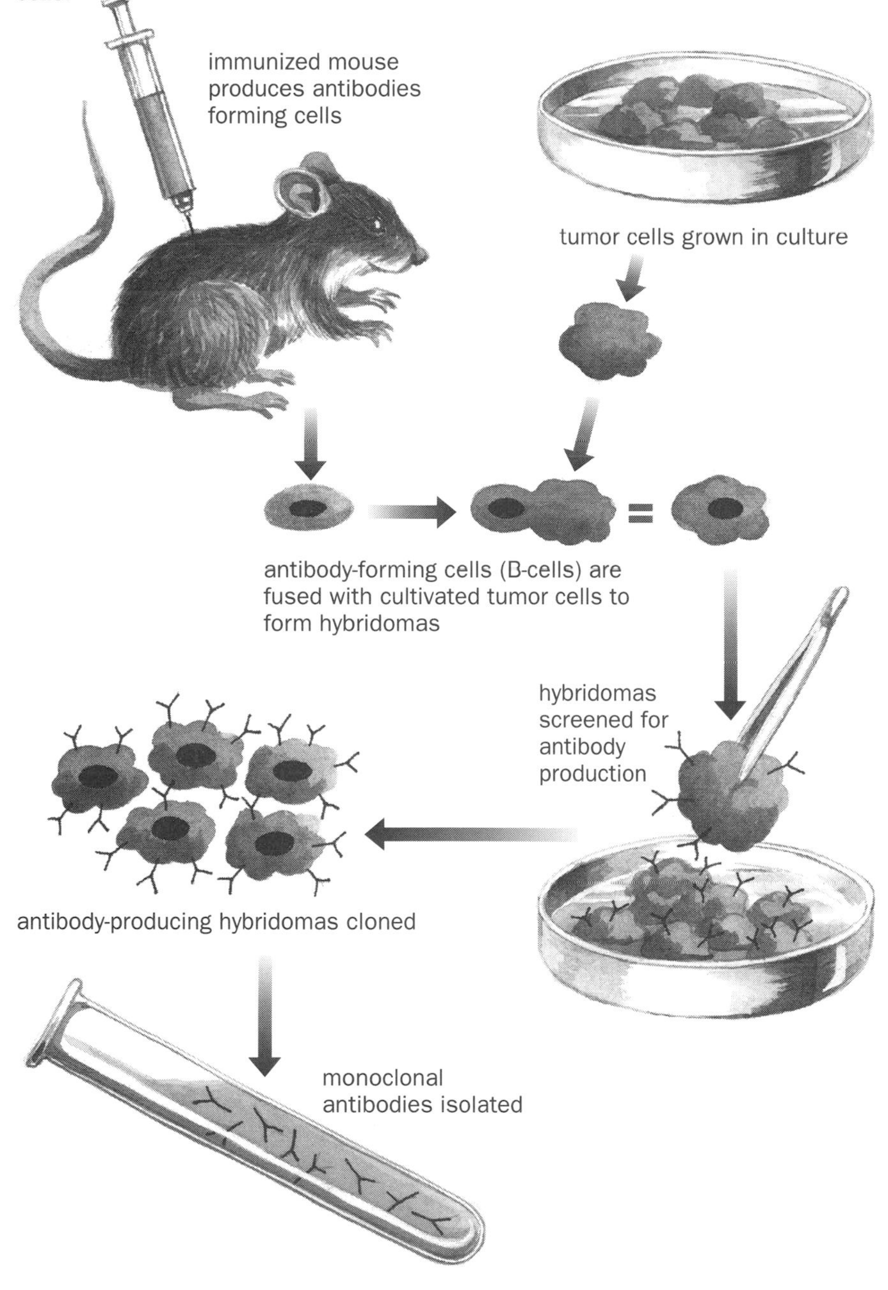

With their highly specific affinity for target substances, mAbs resemble mother seabirds picking out their own offspring in a colony of millions of hungry chicks squawking for attention. Even better, mAbs can locate their quarry even if it moves to different locations and hides among different companions. By 1995, about 70 different mAbs were under development by American companies. Half of these mAbs were aimed at treating or diagnosing various forms of cancer. Consider a few examples of the applications that have made this group of molecules such a winner with the pharmaceutical industry.

Diagnosis: Many diseases are connected with the presence of unusual substances in the body, or with either excessive or very low amounts of normal substances. By adding mAbs to samples of blood or other bodily fluids, then fishing them back out with their targets attached, scientists can get very precise measurements of the amounts of specific substances present. Rapid, sensitive, simple, and accurate, this technique lets doctors diagnose diseases in their very early stages before more obvious symptoms appear.

Treatment: Powerful anticancer drugs can be attached to mAbs that specifically seek cancer cells, allowing them to be carried like guided missiles to their target and avoiding unwanted injury to healthy tissues, and the unpleasant side effects of conventional chemotherapy. A recently published example described the use of cancer-seeking and cancer-destroying radioactive monoclonal antibodies against B-cell lymphoma. The treatment, described as "one of the most promising developments in many years," caused tumors to greatly shrink in size in 70 percent of the patients and to completely disappear in 50 percent within days after a single infusion, with minimal or no side effects. The antibodies were designed to bind to a protein that is found only on the surface of cancerous B-cells and not on any other types of cells.

Monitoring: To follow the progress of a disease or treatment, surgeons can join radioactive isotopes to mAbs, subsequently using detectors to find the location of the radioisotope in the body. For example, this technique can be used to tell surgeons precisely where blood clots or cancer tumors are located.

Autopsies: Pathologists can determine quickly and accurately whether rabies virus, for example, is present in the brain of an animal by adding mAbs that seek out the virus.

Drug purification: Pure chemicals can be cheaply extracted from complex mixtures by adding mAbs that attach only to the chemical molecules wanted. Valuable chemicals present in even a very tiny percentage of the mixture can be separated and purified in this way. For example, this technique helped make interferon widely available.

Screening: To match tissues for transplant operations, doctors can use mAbs to search donor organs for antigens identical to those found in the patient. The process reveals which organs are most similar to the patient's own.

The development and production of mAbs have barely begun to bloom but cutting-edge researchers are already proclaiming the next advance in technology. The first generation of mAbs were simple, unadorned antibodies. The second generation saw mAbs joined to anticancer drugs, radioisotopes, or other useful chemicals. The latest generation are genetically engineered hybrid molecules that combine mAbs with other proteins.

For example, researchers at the U.S. National Cancer Institute and the Weizmann Institute in Israel have engineered a gene to produce a molecule that is a cross between a mAb and a T-cell receptor. The hybrid molecule has binding sites for two different targets: it latches onto a cancer cell and then invites the body's own immune system to kill that cell. When both binding sites do their job, the cancer cell and the immune system cell are connected.

Fusion proteins such as this are being designed to optimize different aspects of treatment, such as targeting diseased cells, binding to cells, penetrating tumor tissue, destroying cells, clearing unwanted matter from the bloodstream, and blocking capillary growth near tumors. This last strategy aims to kill cancer cells indirectly, by cutting off the blood vessels that supply energy and materials for new tumor cell growth.

Designer drugs

The close connection between antibody and antigen is a model for understanding how the body's biochemical systems in general work. One molecule aligns with another and something happens in the body — an infection is blocked, a nerve impulse is sent, a building block of tissue is formed, a cell starts to die. This model is what modern pharmaceutical companies use to design new drugs, custom-made for specific purposes.

"Design" is the right word here, because researchers can now develop drugs on computers before they mess about with test tubes and chemicals. Using three-dimensional images of protein structure, they can rotate their cybermolecules on a screen and study them from all angles to figure out which drugs might best fit active sites on a protein surface. Chemical engineers can manipulate different combinations of drug and disease protein, analyze the functional outcomes of adding one molecule to another, and build up therapeutic molecules from scratch with their computerized construction sets.

In the early days of the biopharmaceutical industry, it was thought that such protein manipulation would provide the ultimate solutions to most health problems. Biological activity, however, is not simply a matter of nuts and bolts. As one professor of chemical engineering has put it: "If you want to make protein, and just make protein, you are in the soybean business."

Living cells are complex, dynamic chemical plants with sophisticated control mechanisms and built-in redundancies. They have multiple pathways for chemical change, sensitive energy needs for chemical reactions to take place, and feedback systems that respond to every change. In this intricate environment, will a drug act only on the proper target and not on something similar? Will it react in the body in the same way it does on the computer screen or in a test tube?

"A drug is a drug is a drug" is not a maxim to be assumed. Researchers have found that the effectiveness of a drug can even depend on how it was made. In one test, for example, antibodies that were produced in living mice had a half-life of 13 hours, while the same antibodies produced in a lab culture of mice cells had a half-life of only three minutes — not long enough for therapeutic use. The difference seems to depend on subtle differences in the cell environment where the antibodies were formed.

Because the therapeutic value of a drug can vary with its environment, a major practical problem in the drug industry is how to mass-produce quality drugs. Physical and chemical conditions inside a small lab flask are not the same as those inside a huge industrial vat. It isn't always possible to predict what may occur chemically when scaling up from lab production to industrial manufacturing, and the molecular properties of a drug can also be altered in small ways, not only during synthesis but also during the recovery and purification processes.

A case study: tryptophan

Tryptophan is a naturally occurring amino acid, used for over 15 years in dietary supplements and infant formulas, and as a treatment for a number of conditions including depression, obesity, and insomnia. In late 1989, it was connected with a sudden outbreak of a debilitating syndrome that resulted in dozens of deaths in the U.S. and caused a variety of adverse effects in as many as 5,000 people.

Immediately following the unexpected epidemic, the U.S. Food and Drug Administration (FDA) recalled the drug and looked for the precise cause of the syndrome. Evidence pointed to contamination, and studies eventually traced the outbreak to supplies of tryptophan produced by a single Japanese manufacturer, Showa Denko.

Tryptophan is manufactured commercially by a bacterial fermentation process, followed by several steps for purification. Between December 1988 and June 1989, Showa Denko changed a number of their manufacturing procedures. They introduced a new, genetically altered strain of bacteria and changed their purification procedures, partially bypassing a step for filtering out impurities. Nearly 60 different impurities were later detected in tryptophan made by Showa Denko, two of which are key suspects in the outbreak of the syndrome. Tryptophan from other manufacturers is free of the contaminants, and continues to be used around the world, both in supplements and clinical research.

The case of the contaminated tryptophan raised concern about how a leading pharmaceutical company can change its manufacturing protocols and not determine that its "new" product contains numerous contaminants. FDA standards for pharmaceutical quality address only the minimum percentage of the labelled product (98.5 percent). The Showa Denko tryptophan exceeded this standard, with a content of 99.65 percent. But the system doesn't examine what types of additional chemicals may be present in the tiny remaining amount of the product.

The case also had fallout that was more political than scientific. The responsibility of the FDA for controlling dietary supplements (as distinct from food and drugs) has been a gray area. In 1991, the Commissioner of Food and Drugs used the tryptophan example as a reason to take a completely new look at how these products should be regulated. The industry countered that this was unnecessary since the epidemic was not the result of dangerous dietary supplements, but rather the lack of regulation of biotechnology.

In 1990, it was revealed that FDA officials delayed informing the public that genetically engineered bacteria were involved in the production of the contaminated tryptophan. They did this at least in part to avoid an adverse impact on the growing biotech industry. The entire incident made many people question the FDA's competence as a regulator.

To some, the news appeared more sinister. Dr. John Fagan is a molecular biologist at the Maharishi University in Iowa and author of a book: *Genetic engineering: the hazards, Vedic engineering: the solutions*. He is opposed to much of biotechnology and claims that the adverse effects of tryptophan in the U.S. in 1989 were a result of genetic engineering.

The production of contaminated and lethal tryptophan by Showa Denko, and the aftermath of the syndrome, are likely to have made many people confused and anxious. The case illustrates how easy it is to blame biotechnology for events that are, in fact, due not to the new technology but to a combination of negligence and lack of adequate regulations and control.

Closing thoughts

A recent newspaper headline reads: "Conquest of disease far from complete." It was a droll understatement, I thought, on a par with: "Sisyphus approaches final stretch of mountain climb." In other words, it illustrates a dangerously false way of thinking about ill health: as a threat we're on our way (albeit slowly) to eliminating once and for all.

The enthusiastic quest to vanquish disease is a common, if unconscious, thread in much of what is written about medical biotechnology. Given the extraordinary progress made in diagnosing and treating many previously untreatable conditions, it's understandable that people expect the trend to continue ever onward and upward. And yet, during the same period of this medical revolution, we've also seen the resurrection of old diseases once thought conquered, such as tuberculosis and polio. There has been an increase in certain cancers, and the advent of some new and even more virulent diseases, such as AIDS and the Ebola virus. Why is that?

To think of disease as defeatable is to think of it as a fixed set of more-or-less inert objects, like bottles on a wall, which researchers with shotguns can pick off one at a time. Ten down, 990 to go. In the more dynamic world of living things, however, the bits of broken glass at the wall keep re-assembling to form new bottles, forever ready to pop back into empty spaces.

Almost as soon as penicillin became widely used, for example, antibiotic-resistant germs began to appear. These resistant strains are simply microbes that our own strategy of defense helps promote. They are mutant survivors of our chemical blitz, passing on their techniques of resistance to subsequent generations to produce new microbes that have escaped antibiotic control. Since microbes have lifetimes measured in minutes or hours rather than years or decades, resistance can take hold and spread very rapidly.

The problem with antibiotics

Alexander Fleming wrote, in 1945, "The greatest possibility of evil in self-medication is the use of too small doses so that instead of clearing up infection, the microbes are educated to resist penicillin..."

The more antibiotics are used, the more resistant the surviving strains become. Those that survive an antibiotic assault that kills 99.9 percent of their fellow germs are more likely to be superstrains, almost invincible to control for a long time to come.

- Throughout Europe in 1979, only 6 percent of pneumococcus strains were resistant to penicillin. In 1989, 44 percent were resistant.
- In the U.S., more than 90 percent of staphylococcus strains are now resistant to penicillin.
- Of the 150 million antibiotic prescriptions written by American doctors each year, it is estimated that almost half are misprescribed or misused.
- Worldwide, an estimated nine billion dollars is wasted each year on ineffective use of antibiotics.
- The spread of resistant germs results in longer hospital stays and use of more expensive alternative drugs.

Take another newspaper story, reporting that a new group of rat-borne diseases is on the rise. The obvious answer is to get rid of rats. But at least one disease, bubonic plague, may paradoxically increase if we follow that approach. Bubonic plague is caused by a bacterium that lives in the rats' blood and is spread by fleas when they go from one animal to another to feed. If we simply get rid of rats, we don't automatically get rid of the bacteria and fleas as well. On the contrary, we create the serious risk that the fleas, when they find the rats suddenly gone, will turn to people for a meal instead, spreading the plague among humans. We could be better off keeping a certain number of rats around.

This type of ecological perspective on disease has also been put forward to explain the apparently sudden appearance of disease-causing agents such as the Ebola virus. One hypothesis proposes that this virus was formerly at home among wild primates and other mammals in tropical forests, where it had achieved an equilibrium with its hosts over many generations. With the removal of forests and the wild species that live in them, some of the more resilient viruses took up residence in people. Accommodating themselves to their new dwelling takes time and trouble, manifested in the violent reactions our bodies have to these unfamiliar tenants.

Recognizing that the struggle to resist one disease or another will always be with us, like death itself, what is the goal of medicine? How does biotechnology help that goal? The overriding approach of biotechnology is to control or fix whatever threatens ill health. It tends to emphasize high-tech intervention and the search for cures.

An alternative model, arguably as effective, sees diseases not so much as things that happen to us that we must get rid of, but largely as the result of unhealthy ways of living. The emphasis is low tech and preventative. In Europe, for example, many water-borne diseases (such as dysentery and typhoid) were dramatically reduced during the early part of the 20th century by the simple move of building better sewers and supplying clean drinking water. No medicines were needed. What was true then is true today. A good diet, exercise, no smoking, clean air and, especially, clean water keep many diseases at bay, while there's no doubt poor nutrition, sedentary lifestyle, smoking, and polluted surroundings encourage ill health.

Ironically, the proliferation of drugs supplied by the medical profession has itself become a health hazard. Most drugs are intended for very specific uses in specific doses, but misuse of such things as sleeping pills and over-the-counter painkillers is a growing problem. A review of nearly 127,000 case records by researchers at the University of Pittsburgh in 1994 found that moderate overdoses of acetaminophen (equivalent to 8 to 20 extra-strength Tylenols taken within 24 hours) can lead to liver damage in people who have been eating very little or drinking alcohol. Another study found that heavy use of painkillers was associated with as many as 5,000 cases of kidney failure in the United States each year.

With growing pressure to cut health care costs, more studies are beginning to question the expensive, high-tech approach of the past few decades. Victims of low back pain, for example, which includes a surprising three-quarters of American adults under age 50, cost the United States nearly $20 billion a year in direct medical expenses. In 1995, analysts from the U.S. Agency for Health Care Policy and Research recommended against the use of some routine treatments including surgery, finding in many cases that the pain would go after a month of moderate exercise, Aspirin, and spinal manipulation.

There is no doubt that biotechnology has vastly increased our scientific understanding of the body and the way it works at the most fundamental level. We know far more than ever before about the origins and development of diseases and have fast and accurate tools for detecting and diagnosing almost every kind of illness. What is less clear are the economic and social results of this revolution in knowledge.

Despite its promise of cures, biotechnology has had more impact so far on screening and diagnosis than on therapy. Doctors are increasingly able to use genetic screening and other tests to reveal the likelihood of someone getting a disease, without having the treatment for it. This raises the issue of the technology's true value as well as ethical dilemmas. Disclosing a patient's chances of getting a particular disease in the future can raise anxieties in patients and their families for years over something that may never happen, and can affect their relationships, employment, and health insurance.

Are we tempted to use the latest technology just because it is the latest, believing without question that it offers an improvement over older ways? Like children with too many toys, overburdened physicians may pick up and discard one approach after another, so distracted by novelty, or lured by the appearance of a more powerful tool, that questions of suitability and effectiveness are left for later.

Rather than applauding every new breakthrough in knowledge, and welcoming every new application of biotechnology as an undoubted good for society, we need more than ever to look at the priorities of our finite health care systems. Many health problems might be solved more effectively by a change in economic and social conditions than by advanced medical technology and drugs. For example, statistics show that poorer people get sick more frequently and die younger for a variety of reasons that include inadequate housing and nutrition, lack of education, drug addiction, living in polluted environments, and working in hazardous jobs.

Expensive, sophisticated, high-tech medicine does

not guarantee longer, healthier lives any more than expensive, sophisticated, high-tech weapons guarantee world peace. The United States deploys about 14 percent of its total economy to maintain the world's most technologically advanced medical system. Yet American life expectancy at birth ranks behind that of 15 other nations, all of which spend proportionately far less on health care. The infant mortality rate in the United States is higher than in 21 other countries, and black babies are more than twice as likely to die in infancy as white babies.

The American experience should warn us not to be dazzled by the mere firepower of medical technology. The triumphs of biotechnology are saving some individuals from some diseases. They give us exciting and finely detailed molecular descriptions of what we are. But they cannot, in the end, save us from who we are.

Chapter 4
Biotechnology on the Farm

Is biotechnology the answer to a hungry world's food supply, or is that only self-serving rhetoric from a profit-hungry agribusiness? You hear both sides in the talk over biotech on the farm and it's not always easy to sort out the facts from the propaganda. Where food is concerned, people are especially wary. Will genetically engineered foods cause health problems? How does changing technology affect farmers? What impacts will it have on the environment? As the first products of agricultural biotechnology reach supermarket shelves, the debates seem likely to continue for some time.

There's no escaping the fact that agriculture is BIG business. The production and marketing of food and other farm products, such as cotton and tobacco, make up the world's largest single industrial sector. Applications of biotechnology in agriculture rank second only to those in medicine, and estimates of the potential market range as high as $67 billion per year as the 21st century begins. It's no surprise that companies such as Monsanto — a major agricultural biotechnology corporation with

30,000 employees and net sales of almost $10 billion per year — predict a future for farming that will be more efficient, reliable, environmentally friendly, and profitable.

The sorts of changes that biotech companies expect to bring to agriculture in the next few years are unlikely to make much difference in the look or even, in most cases, the taste of the food you'll find in stores. But engineered varieties of plants and animals will have a big impact on the business of farming. The focus of most crop research is to increase yields, using biotechnology to develop plants that:

- survive drought, frost, and other environmental stresses
- resist insect pests and diseases
- tolerate herbicides, allowing farmers to spray weed-killer on fields without damaging their crops.

Companies are also looking at ways to develop crops that have less need for artificial fertilizers, and to produce grains, fruits, and vegetables with characteristics such as a longer shelf life and a modified oil composition or nutritional content.

For livestock, the goals of biotechnology are to improve the quality of meat, milk, eggs, and wool, and to produce healthier and faster-growing animals. Some of the same techniques used in human medicine are used to develop vaccines, diagnoses, and disease-resistance for farm animals, while reproductive technologies allow farmers to get more offspring from select animals. Fine-tuning genetic control can give us such things as designer milk (custom-made for yogurt production or lactose-intolerant consumers) and flocks of engineered sheep growing the ideal wool for carpet-making.

Will biotechnology be able to follow through on such promises? While earlier technologies have managed to nearly triple grain yields since 1950, increased production has its price. The use of pesticides, fertilizers, and irrigation has grown sharply over this period, causing damage to the environment and human health. And many of the gains made are beginning to slip. For example, insecticide use in the United States increased tenfold between 1945 and 1989, but crop losses to insects in the same period nearly doubled. Critics of the industry point to outcomes like this to argue that corporate farming is not sustainable, and that our entire approach to food production must be reconsidered. It's still not clear whether biotechnology can satisfy both camps, maintaining food output while decreasing environmental risks.

One thing that can safely be predicted is that the farmer's twin banes of weather and economic conditions will continue to affect whatever means of production we choose. Puncturing the inflated optimism of industry with the sharp point of nature's vicissitude is this comment from the Worldwatch Institute's *State of the World 1990* report.

> "Grandiose claims about biotechnology and food production have been common since the first successful attempts at genetic engineering in the early seventies. As recently as 1984, one writer predicted that 'in 5 to 10 years, Saudi Arabia may look like the wheat fields of Kansas.' The unfortunate reality in 1989 — when Kansas lost over a third of its wheat crop to drought — was that the wheat fields of Kansas came to resemble the still fallow Saudi Arabian desert."

Milking it for all it's worth

One of the first genetically engineered products available to farmers was bovine somatotropin (BST), also called bovine growth hormone (BGH). Made naturally in the pituitary gland of cattle, the hormone promotes growth in calves and regulates milk production in mature dairy cows. The engineered version of the hormone is used to increase a cow's milk yield by up to 20 percent. It is manufactured by bacteria using copies of the cow's genes, so the product administered to the cow is essentially the same as that made by the cow herself. Nonetheless, the use of recombinant BST has been the subject of heated controversy since the mid-1980s, and the battle for public opinion still continues.

At this time, there are at least 16,000 references to BST on the Internet. Although the hormone has been approved for commercial use by American farmers since 1994, it still concerns many people. A review of this long-running debate helps illustrate some of the issues raised by agricultural biotechnology as a whole.

First of all, why give a cow something she already has? The short answer is economics. A dairy cow's normal milk production follows a cycle, rising to a peak about 50 days after calving then steadily declining over the following 10 months. The extra energy she needs to make milk comes initially from her body fat, but after the peak she must eat more food to meet the ongoing demands of lactation. Dairy farmers use BST during this second half of the cycle to boost milk production, increasing the yield per cow without increasing their feeding costs.

Much of the pressure on farmers to increase the yield

per cow comes from government-established quotas on milk production, both in Europe and North America. Faced with controlled milk volumes and prices, the farmers' only way to maintain income in the face of inflation is to reduce costs. In other words, BST is used to meet economic policies, not consumer demand. Far from needing more milk for the market, the main problem of dairy-producing countries during the 1970s and 1980s was overproduction. The manufactured hormone seems to offer a political and economic way out of subsidizing dairy herds, allowing farmers to produce almost as much milk from fewer cows.

The overall economic effect of BST on farmers may depend on their scale of operation. The added costs of medication and veterinary visits associated with use of the drug, and the need for rigorous attention to details of feeding and housing, make it uneconomical to use on a small scale. Some observers worry that widespread adoption of the hormone might put small and medium dairy farms with fewer than 50 cows out of business.

Outspoken corporate critic and food-system analyst Brewster Kneen sees BST as another step on a technological treadmill that confines family farmers to a narrow role in a capital-intensive food production system. The goals of this system, Kneen says, are not so much to produce good-quality food efficiently as to maximize profit for the manufacturers and sellers of technology and information. In the long run, he argues, family farms might be a better bet for sustainable food production than the shorter-term mandate of agribusiness shareholders.

Aside from the economic issue of survival for individual dairy farmers, and the broad question of who benefits from BST use other than the manufacturers, the main focus of public concern has been on animal and consumer welfare.

Is it cruel and unhealthy to force cows to produce more milk? With each shipment of Posilac (the name of Monsanto's commerical BST) comes a packing label warning of possible side effects. These range from swelling at the site of injection, to indigestion and diarrhea, decreased appetite, and a reduced rate of pregnancy. Opponents of the drug claim it also produces brittle bones, lameness, mastitis (inflammation of the mammary glands), and lowered resistance to disease, so that cows receiving the hormone must be treated with more medications, including antibiotics.

During the first year after Posilac was approved for use in the United States on February 1, 1994, 14 million doses were administered on 13,000 farms, representing 11 percent of American milk producers. Close monitoring by the U.S. Food and Drug Administration between February and August 1995 cited 509 "adverse reactions" in herds given Posilac, but all of them involved conditions also found in herds that are not on BST. Given this fact, biotech advocates, including farmers who use Posilac, argue that health problems are caused not by the hormone but by inadequate management and improper feeding, hygiene, or veterinary care.

Animal welfare organizations oppose the use of BST on ethical grounds. Cows on BST require more medical attention, including injections. The purpose of this extra treatment is not therapeutic, or even prophylactic, but designed only to increase milk supply. Some object that it is unethical to view animals, which are capable of suffering, as food-producing machines. This argument extends to livestock production in general.

Does milk from BST-treated cows pose a health risk to people who drink it? In announcing the approval of BST in 1994, FDA Commissioner David Kessler said, "This has been one of the most extensively studied animal drug products to be reviewed by the agency. The public can be confident that milk and meat from BST-treated cows is safe to consumers." Yet anxiety about drinking milk from such cows is a major concern of the product's opponents.

The fear is that milk from BST-treated cows contains high levels of hormones, which could be a health risk to humans, especially children. One substance often raised in the debate is insulin-like growth factor (IGF), a chemical that stimulates the growth of cells in the infant's gut. IGF is normally found in both human and cow milk just after the young are born. Most published reports find that IGF levels are no higher in milk from BST-treated cows than from untreated cows. People also worry that BST-treated cows have more antibiotics in their milk, but it has not yet been shown if this is the case.

Caught between industry demands and public anxiety, many governments opted for extreme caution over BST. Like any new product seeking approval for public use, it has had to undergo lengthy testing to determine its safety, quality, purity, and stability. Trials to measure its effects on human and animal health involve much larger doses than anything people can reasonably expect exposure to.

In 1994, the U.S. National Academy of Sciences Board on Agriculture reviewed studies of BST from more than 20 different scientific sources and publications, including the U.S. Federal Department of Agriculture, the National Institutes of Health, American Academy of

Pediatrics, World Health Organization, Food and Agriculture Organization of the United Nations, EEC Committee for Veterinary Medicinal Products, and the *Journal of the American Medical Association*. The chair of the board, Dr. Dale E. Baumann, concluded in his report: "There are areas of biology in which knowledgeable experts disagree, but safety of foods from BST-treated animals is not one of them." The studies showed unanimously that "composition and nutritional value of milk from BST-treated cows is essentially the same as that of milk from untreated cows."

By 1996, some 15 countries worldwide had licensed the use of BST. The European Union placed a moratorium on its use until the year 2000 and banned the import of milk from BST-treated cows, but cited the oversupply of milk as the reason, rather than any concerns about safety. Canada has still not approved use of the hormone, pending reviews of its impact on the dairy industry, among other considerations.

In the absence of any clear evidence of harm to health, and with the growing acceptance of the drug in more countries, opponents of BST shifted the battle to the question of labeling milk sold in stores. The issue became one of choice. Those who still doubted official reassurances about safety should at least have the information to let them choose whether or not to buy milk from BST-treated cows. Countering this argument, the Food and Nutrition Science Alliance (representing the American Institute of Nutrition, American Society for Clinical Nutrition, Institute of Food Technologies, and American Dietetic Association) responded that, since milk from BST-treated cows is not different from other milk and poses no demonstrated health risks, labels on dairy products claiming to be BST-free would be "neither meaningful" nor "verifiable."

But the matter isn't settled, and criticisms of the hormone continue to appear in the media and at public debates. If this uncertainty confuses you, you're not alone. Public attitudes to biotechnology depend on whom people decide to believe. A survey carried out by the Canadian Institute of Biotechnology in September 1993 showed that doctors and nutritionists are at the top of the public's list for trustworthy information, while government regulators, media, and industry officials are at the bottom.

Ironically, further developments in genetic engineering may move the long-running debate over bovine growth hormone into new pastures. Cows could eventually be genetically altered so they produce more BST themselves, eliminating the need to inject the synthetic hormone.

Let us spray

To increase the world's food supply, farmers must not only grow more food but also reduce food losses. As much as a third of the world's crops are lost to pests and diseases, so protecting plants from these hazards is one of agriculture's biggest challenges.

The word "pests" typically conjures up images of insects nibbling holes in fruits and vegetables; however, farmers face another problem that's less obvious but just as troublesome. About 60 percent of the chemicals used to control pests in the United States today are directed against weeds — wild plants that thrive and spread in the same conditions as crops. There are about 30,000 different kinds of weeds worldwide, competing with crop seedlings for space, water, light, and nutrients, and significantly reducing yields.

Insect facts

Most insects are harmless or, like honey bees, even helpful to farmers. Only about one percent of the world's million or so insect species cause problems for people and only about 100 species seriously damage crops.

The difficulty with controlling weeds chemically is that herbicides (weedkillers) can damage crop plants as well as weeds. Thus, the quantity and strength of the herbicide that a farmer can use is limited by the sensitivity of the crops. Developing herbicide-tolerant crops has been a major priority for agricultural biotech companies. In Canada, for example, almost 86 percent (or about 1,600) of all tests of genetically altered plants between 1988 and 1995 involved herbicide tolerance.

Clues to ways of developing herbicide tolerance came from studies of mutated weeds that had become naturally resistant to the herbicide triazine in farmers' fields. At least 55 species of weeds are now resistant to the triazine group of herbicides. Their resistance is due to a change in a single amino acid in a protein that the herbicide attacks — a simple kind of mutation that can easily be induced by genetic engineering. The great advantage of producing herbicide-tolerant plants by genetic engineering, rather than by traditional selective breeding, is that genes for tolerance can be cloned and transferred into a number of different crops.

One of the first successful examples of engineered resistance was against a widely used herbicide named glyphosate, marketed by Monsanto under the name Roundup. Glyphosate works by binding onto and inactivating an enzyme plants need to synthesize amino acids. The gene coding for this enzyme was altered to produce a modified enzyme that had less affinity for the herbicide but still retained its function. By 1996, the gene for herbicide resistance had been introduced to over 30 crop and forest plants, enabling them to withstand otherwise lethal doses of glyphosate.

The attraction of herbicide-tolerant crops for farmers is that it lets them control weeds more efficiently and cheaply. Freed to use a single, effective spray without harming their crops, they need fewer applications of herbicide. This saves time in the fields, lowers the costs of fuel and chemicals, and reduces the farmers' exposure to herbicides. In addition, less herbicide use means less environmental damage.

If herbicide-tolerance cuts crop losses, it should be welcome. But suspicion of big corporations and their priorities raises doubts. On the economic front, it is argued that these developments benefit the seed and chemical producers more than anyone else, shackling farmers to dependence on particular crop varieties and sprays. It's worth noting that many of the chemical companies that manufacture herbicides also own major seed companies, and stand to gain by selling both seeds for herbicide-resistant crops and the herbicides that help control weeds.

Will herbicide-resistant crops need less spraying? Some opponents worry that the opposite might happen. Farmers might use more herbicides to control weeds once they start growing crops that are immune to spray damage, adding to the already heavy load of chemical contamination in soil and groundwater. About 275 million kg (606 million lb) of herbicides are applied to crops in the United States alone every year. And while intended to kill weeds, herbicides such as glyphosate are also toxic to such organisms as spiders, earthworms, and fish.

Also on the environmental front, there is the concern that engineered genes for herbicide resistance may spread from crop plants to wild species through cross-pollination. This could produce resistant weeds that survive spraying, setting the struggle for weed control back to the beginning.

The strategy of attacking weeds with chemicals may be simple and profitable, but it concentrates on the symptoms of the weed problem rather than the causes. As a result, the industry's focus on producing herbicide-tolerant crops could undermine efforts to encourage alternative, nonchemical methods of weed control that may be more sustainable in the long term.

Integrated pest management (IPM) is an approach that combines reduced use of chemicals with alternative strategies such as crop rotation, intercropping (growing two or more compatible crops in the same field), and control by predators. Compared with the amount of money and research devoted to high-tech methods of pest control (it takes on average 10 years and between $20 million and $45 million to develop a new pesticide), IPM receives little attention from either corporations or the government. The total federal funding for IPM in the United States in 1986, for example, was less than $20 million — one-tenth of one percent of the $26 billion paid to farmers in crop subsidies the same year.

Developing regulations

Since 1986 there have been over 2,000 field trials of transgenic crops around the world, exposing natural ecosystems to the introduction of engineered genes. But while genetically novel organisms establish their place in agriculture, regulations governing their use are inconsistent. Proponents of biotechnology support deregulation, a trend being followed by the United States. Others are concerned about the dangers of releasing genetically engineered organisms before their safety has been assessed.

In spring 1995, canola seeds genetically modified for herbicide resistance were the first crops to be approved for commercial planting in Canada, and applications for similarly altered corn and flax were in the works in 1996. The canola were grown for their oil, but oil from the engineered crops couldn't be sold to the United States without approval from the U.S. Food and Drug Administration. And although the crops could be grown by farmers in Canada, they were not initially approved for use as livestock feed.

Canada leads the world in the development of herbicide-resistant crops, and its criteria for approving new varieties for the market are strict. One transgenic variety of canola was failed because its protein content was below the minimum allowable by less than one percent. The company that developed the new canola argued that protein content was irrelevant since most of the crop was intended for producing oil, not feed, but the regulatory committee narrowly voted not to approve it for commercial use.

As the first large-scale plantings of engineered crops begin to establish track records for safety, farming countries around the world are being forced to develop regulations to deal with them. And every year, there are more and more transgenic crops to regulate. The vice-president of the Canadian Seed Trade Association predicts that in a few years, novel traits produced by genetic engineering will be found in the *majority* of new varieties being submitted for approval. To avoid swamping the system in a backlog of delays, the pressure is on to develop standard international requirements. In the end, there seems little doubt that engineered crops and livestock will play a growing role in food production. The final extent and effect of that role, and the social and environmental costs of this new direction in farming, however, remain open questions.

Great expectations

Between 1950 and 1984, world grain output rose an astonishing 260 percent, thanks to a combination of improved varieties, irrigation, artificial fertilizers, and chemical pest control. During the same period, the number of people on the planet almost doubled. Today, world population growth adds about 90 million new mouths to feed every year, while land degradation, pest resistance, pollution, and climate changes have slowed or leveled growth in crop production. In the early 1990s, world grain production per capita began to decline for the first time since the Second World War. There are many who believe that biotechnology may now be the only way to reverse this problematic new trend and maintain food supplies (Figure 4.1).

Genetic engineering can be used to modify different stages of crop production, from speeding up early growth

Figure 4.1

Per capita grain production

By the mid-1990s, world grain production slipped below 300 kilograms per person for the first time since 1970.

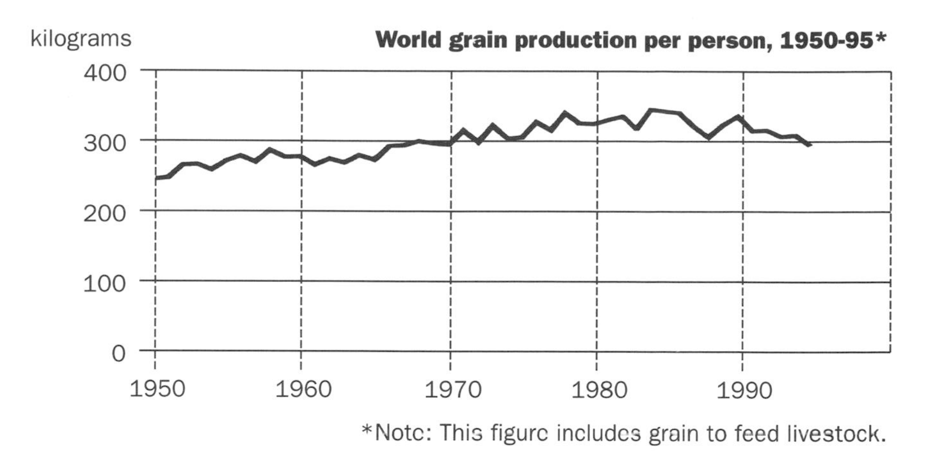

of food plants, to increasing yields, to slowing down ripening or wilting. Since the form and function of a plant depend a great deal on its genes, the ultimate hope is to engineer optimal plants for every growing condition and market niche.

A little plant with a big future

Much of what we know about the molecular basis of plant growth and development has come in the past few years from a plant equivalent of the Human Genome Project. The *Arabidopsis thaliana* Genome Research Project was established in 1990 to uncover the genetic details of a small plant in the mustard family. Scientists from 35 nations have already produced detailed genetic maps of *Arabidopsis thaliana*, chosen for this massive research effort because it is small, has a short life cycle, is a prolific seed producer, and has the smallest known genome of any flowering plant — only 100 million base pairs. What scientists learn about the biology of this plant may be applied to many other plants.

Knowledge of how even a single gene works can have a broad impact. Most herbicide resistance, for example, is based on single gene alteration. The simple basis of this trait is probably another reason why it was one of the first to be successfully manipulated. Researchers have also discovered that a single gene helps tomatoes resist bacterial disease. The recent identification and sequencing of this gene gives scientists the first clear glimpse of a genetically based disease-resistance mechanism, offering clues for identifying similar genes in other plants. Engineering plants to withstand diseases and pests without the need for chemical sprays could be a way to boost food production without adding more stress to the environment.

Pests and diseases

Take a field of rye and watch what happens when the plants are infected with a fungus. The fungus sweeps through the field and plants die. But not all of them. The question is: why does one plant succumb to a fungal disease while a second plant growing near it resists? Researchers have found that, in many cases, the difference between resistance and susceptibility is simply the rate of the plant's response. If a plant can respond to the first attack of fungi rapidly, then it can resist further damage.

Using this knowledge, farmers can now inoculate their crops against some fungal diseases. Slow-responding plants are given a head start by being deliberately infected with disarmed fungi — the same principle used to vaccinate children against infectious diseases. The plants' response to these harmless versions prepares them to resist virulent forms that may arrive later.

Crops have also been inoculated against viral diseases by giving them copies of viral genes that limit the ability of invading viruses to replicate inside the plant cells. This approach induces resistance to diseases that were previously unmanageable, and opens new possibilities for effectively controlling viral diseases before they become established.

Fungal diseases of fruits, vegetables, and grains alone can cost growers billions of dollars annually. New fungal-resisting genes can now be inserted into corn using a gene gun — an instrument that literally shoots tiny bullets of microscopic metal particles coated with genes. It shoots genes into clusters of cells, which are then stimulated to multiply and grow into complete plants.

Figure 4.2 A researcher uses a gene gun to introduce new genes into an organism. DNA is coated onto microscopic gold or tungsten pellets that are propelled by the particle gun into plant or animal tissues that are in a petri dish (inset).

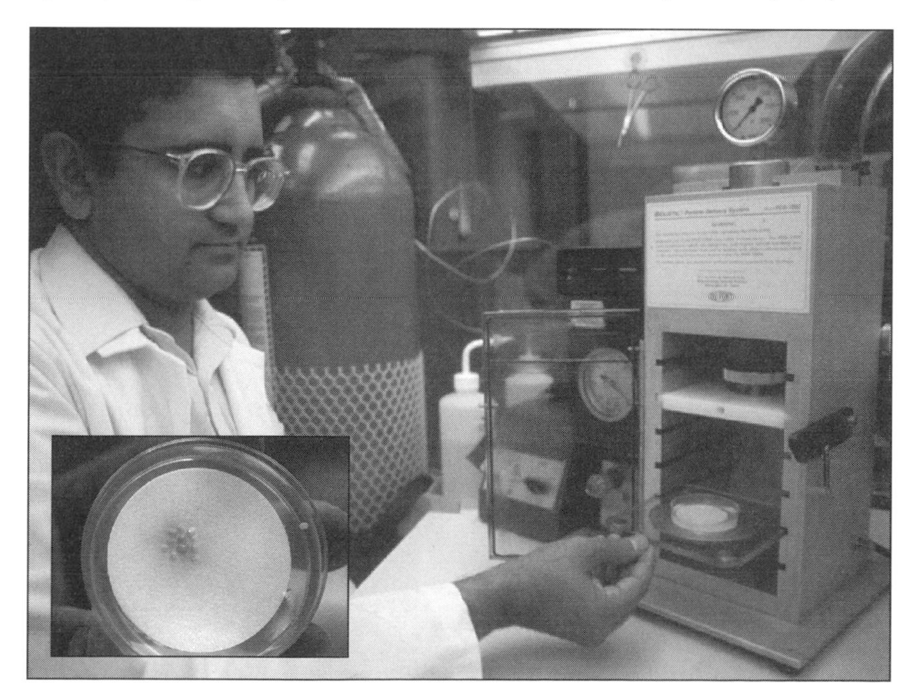

Traditionally, disease resistance was developed in crop strains through selective breeding of naturally resistant individuals. The process is now made much faster by cloning the genes responsible for resistance and inserting them into other plants, cutting the time needed to develop new strains from about 12 years to only two or three years. Once a resistant strain is established, the genes will persist in future generations through normal breeding methods. This technique has been used to culture barley plants with resistance to yellow dwarf virus.

On paper, these strategies tend to sound foolproof. But the tactics used to combat diseases must be as sophisticated and flexible as nature itself if they are to succeed for very long. Viruses, in particular, can quickly develop mutants that sidestep host resistance based on a single gene. To avoid this potential failure of their work, scientists plan to insert different genes for viral resistance into different plants and then crossbreed them so that offspring have more than one route of resistance and viruses will have a harder time evolving ways around their varied defenses. This approach combines the latest in biotechnology with the much older and still very valuable technique of crossbreeding to maintain genetic diversity. It is a common error to suppose that genetic engineering will replace the need for plant breeders.

One of the lessons taught by the widespread pest spraying programs of the '50s and '60s is that simplistic approaches to controlling pests or diseases don't last. Insect pests, with their rapid rates of reproduction, can quickly evolve resistance to toxic sprays while the buildup of the same poisons causes populations of their natural predators to decline. The result is a rebound of organisms that are much harder to get rid of. For all the power offered by biotechnology, ultimate success will still depend on the degree to which we understand the natural systems we want to manipulate. Co-opting nature is a better idea than opposing it. To that end, many agricultural scientists are investigating the potential of biological control methods.

Since the beginning of the century, the U.S. has imported and released approximately 800 natural enemies of insect pests, and about 40 percent of these continue to provide some level of pest control. How can biotechnology help give these predators and diseases of pests a helping hand?

When nibbled by insects, many plants release chemicals that drift through the air. Some predatory insects that feed on the plant nibblers use these chemical signals to zoom in on infested plants for an insect meal. For example, corn leaves chewed by beet armyworm caterpillars put out volatile compounds that draw the attention of parasitic wasps, which lay their eggs in the caterpillars. The wasps are very discriminating, ignoring similar odors released when leaves are damaged mechanically, for example by mowing. By analyzing the chemical molecules that predatory insects use to find their prey, researchers could engineer crops to produce stronger signals when attacked, attracting higher populations of the pests' natural enemies.

Instead of using chemicals to call in the cavalry, other plants use them directly for defense. Many secrete their unpleasant chemicals through forests of tiny, hollow hairs covering their leaves and stems. If you've ever grabbed a stinging nettle, you'll know how this works. Plant strategies vary. Some plant hairs produce sticky sugars that act like flypapers, trapping insects landing on the plant. Others, including citrus plants, tomatoes, and aromatic herbs like sage, thyme, and mint, produce chemicals that either repel or poison insects. Decoding the genetic control of these protective chemicals could potentially let bioengineers alter the quantity or type of chemicals produced, or add a built-in defense to other plants. These natural defenses, boosted by genetic engineering, may be all the pest protection that some plants need.

As well as causing billions of dollars' worth of direct damage to crops by feeding on them, many sucking insect pests, such as whiteflies, aphids, and leafhoppers, transmit viruses and bacteria that cause devastating plant diseases. One new approach to controlling these particular pests depends on the fact that sucking insects have vital symbiotic bacteria in their bodies. The bacteria provide essential amino acids to their insect hosts, benefiting them in the same sort of way that symbiotic bacteria in the stomachs of cows help the cows digest grass. By exploring ways to inactivate these little-studied microbes, either by manipulating their genes or by engineering an antimicrobial agent into the plants the insects feed on, scientists would have a powerful way of indirectly controlling the pests.

A versatile bacterium

One of the most successful agents of biological control, first discovered in the early 1980s, is *Bacillus thuringiensis* (Bt), a bacterium that makes insecticidal chemicals. When ingested by insects, the bacterial spores germinate and produce their toxins, eventually killing the insect as part of their own life cycle. Different strains of the bacterium make their own toxins, each of which has its own range

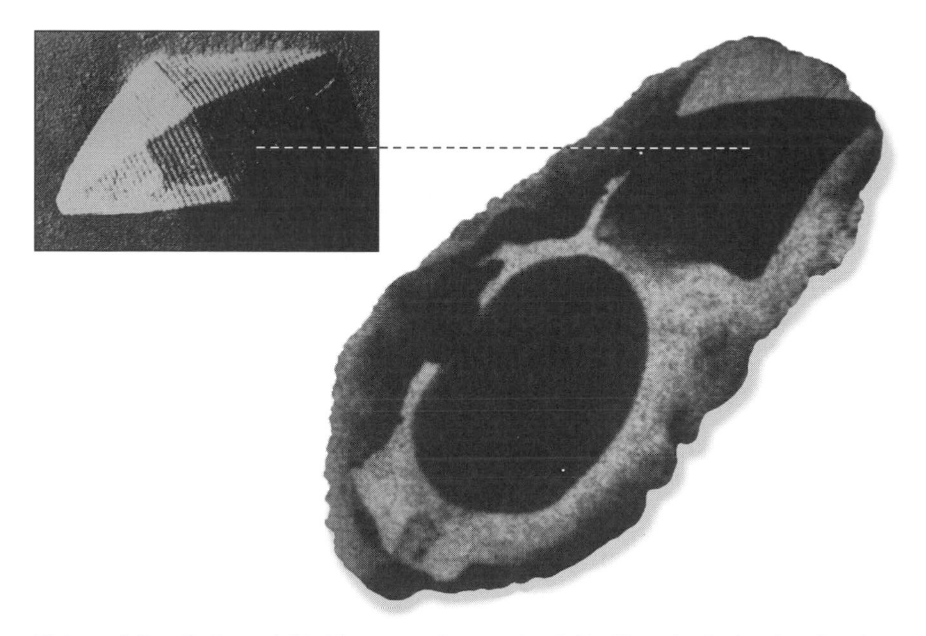

Figure 4.3 Bottom right: Electron micrograph of *Bacillus thuringiensis*, showing its bipyramidal toxic crystal. Top left: A released toxic crystal.

of insect targets. To date, scientists have discovered the genetic coding for over 50 Bt insecticides, and the chemicals are being used as controls against several destructive pests such as gyspy moth caterpillars, tobacco hornworms, Colorado potato beetles, and cotton bollworms.

The Bt insecticide needs an alkaline environment to produce its effects, the kind of environment found in the midguts of many leaf-eating insects. On being activated, the protein binds to cell membranes in the gut, affecting the insect's ability to regulate osmotic pressure and eventually killing it from massive water uptake. In the highly acidic conditions of most mammalian guts, Bt is quickly broken down into harmless chemicals.

Bt has several advantages over conventional pesticides. The bacterial toxins are specific to a few target species of insects and safe for other species. They are quickly denatured by ultraviolet light, and so don't persist to pollute soil and water, or to work their way through food chains.

Pesticide sprays containing dead Bt have been widely used in Canada to control certain moth caterpillars that damage forests. The toxin in the dead bacteria remains active for a brief period before being broken down by sunlight, and using dead microorganisms avoids the spread of live bacteria in the environment.

Bacteria themselves are used in sprays because it is costly and difficult to synthesize the bacterial toxin in commercial quantities. But genetic engineering can now be used to modify the bacteria, increasing their growth rate or altering the quantity, strength, and specificity of the toxin they produce. These kinds of modifications increase the effectiveness of the Bt spray and decrease the number of applications needed, lowering the overall costs of pest control.

A second method of pest control based on Bt cuts out the need for spraying by inserting the genes for making Bt toxin directly into plants. Bt genes have already been inserted into tomatoes, tobacco, corn, cotton, and potatoes to produce pest-resistant varieties. In May 1995, NewLeaf Russet Burbank potatoes became the first genetically modified, insect-resistant crop to receive full U.S. federal regulatory approval for commercialization, and grocery stores now sell potatoes with added bacterial genes. Bt corn and Bt cotton have also been approved and commercially grown.

While some people balk at the very notion of transgenic foods, evidence supports the view that this approach to protecting crops from pests is safer for human and animal health and the environment than the use of synthetic chemical pesticides. A more serious concern over Bt is whether its widespread use on different crops over enormous areas of farmland will simply speed up the development of pest resistance, making a relatively safe and useful pesticide useless.

One reason Bt-engineered crops are expected to promote pest resistance is that they produce the toxin continuously, unlike Bt sprays, which expose insects only periodically. By mid-1996, two species of insect pests had evolved resistance to Bt in the field, and another 10 species had shown evidence in the lab of being able to evolve resistance. Tobacco budworm, for example, a destructive pest of cotton, developed a strain in the lab able to resist 5,000 times more Bt toxin than it takes to kill nonresistant strains.

Can anything be done to prevent resistance from

Figure 4.4 The transgenic cotton (left) has been engineered to produce much larger bolls than regular cotton (right).

developing in the wild and undermining the effectiveness of Bt? An important lesson learned from the past is that no method of pest control should be considered alone and in isolation from the environment. The key to longer-term success of Bt is to use it in combination with other strategies, an approach that is increasingly being mandated as part of the government approval of new products and the manufacturer's terms of sale to crop producers.

Products such as Bt are not sold indiscriminately to anyone who wants them, to use as they please. Detailed contracts of sale and licences for use set out strict conditions that farmers must adhere to — conditions designed to prevent the kind of overuse that promotes pest resistance. In North America, investigators from government agencies and from the manufacturers check to ensure that Bt products are being used by farmers in the proper way. A typical customer for Bt-expressing crops is either encouraged or, in some cases, required to rotate the genetically altered crop with others, to mix different varieties of seeds, to continue using some chemical sprays where needed, and to plant areas of crops that don't contain Bt toxins among engineered plants that do.

This last strategy provides refuges where nonresistant pests can survive. These susceptible insects crossbreed with their resistant neighbors, reducing the chance of resistant genes spreading quickly through the population. Susceptibility is usually a dominant trait, while resistance to insecticides is often produced by recessive genes. The hybrid offspring of resistant and susceptible insects will be susceptible to Bt toxins, and the scheme works well if there are enough refuges in the right locations.

The small loss of crops caused by allowing some pests to survive is a tiny price to pay compared with the long-term cost of chasing new generations of resistant pests with one chemical after another. Integrated pest management (IPM) is a much more sophisticated approach than used in the past, where field upon field was grown with the same crop and sprayed with the same spray year after year.

Bt's advantages, however, may quickly prove to be its downfall if more farmers rush to plant Bt-engineered crops. An early warning sign came in the summer of 1996, when thousands of acres of one of the first Bt cotton crops grown in the southern U.S. were infested by cotton bollworms. Although less than one percent of the two million acres planted with Bt cotton was affected, this outcome alerted critics and took some shine off Bt's glowing promise.

The damage may have been the result of an overall increase in pests due to an increase in corn crops planted in the region. (Corn is another host of the cotton bollworm.) Another possibility is that levels of toxin expressed in the crop weren't high enough to kill most of the bollworms. Until we know more about it, say some environmental groups, the government should place a moratorium on further planting of Bt crops. Whatever the final analysis, the infestation of engineered crops is a valuable reminder that 100 percent elimination of pests is neither possible nor desirable. The best we can do is manage the equilibrium to our advantage for a time.

Weather and soil

Everything that biotechnology can do to increase yields and decrease weeds, animal pests, and diseases can quickly be undone by the farmer's enduring nemesis —

bad weather. Although better forecasts help farmers plan for minor problems (such as unseasonable frosts or rain), a major flood, snowstorm, drought, or high winds can knock billions of dollars off the potential value of a year's planting.

Today's farmer isn't quite as much at the mercy of the weather as yesterday's, however. Large, automatically controlled greenhouses help protect high-value crops such as tomatoes, cucumbers, and peppers. And for field crops, genetic engineering can shift odds in the farmer's favor by improving a plant's tolerance of drought, heat stress, frost, or wind damage. Learning how plants tolerate cold may even make it possible to modify subtropical plants so they can be grown in cooler climates.

In Canada, scientists are testing genetically modified alfalfa, grapes, and winter barley for improved freezing tolerance. Even a small change can bring big benefits. For example, researchers estimate that grape production in southern Ontario could double by developing grape varieties able to withstand freezing at temperatures 2°C lower than the minimum endured by current vines.

Many of the physiological results of stress, in people as well as plants, are produced by a destructive form of oxygen called oxygen free radicals. These activated molecules increase in plants subjected to freezing, drying, flooding, or disease. They disrupt cell structure by attacking proteins, fats, and nucleic acids. Plants already have some defenses against this type of damage in the form of enzymes and vitamins, which act as antioxidants. To boost this natural defense system in alfalfa, researchers are attempting to increase the plants' production of the protective enzymes by genetic engineering.

A method of resisting frost damage already in use

employs genetically engineered bacteria. The bacteria, which live on plant leaves, have protein coats that stimulate ice crystals to form around them when air temperatures drop below freezing point. The ice crystals rupture plant cells, releasing nutrients, which the bacteria use for their growth. In the absence of these bacteria, frost does not form so readily on the leaf surface and the plants can survive without damage at lower temperatures.

Since it would be practically impossible to remove bacteria from the leaves, scientists have instead made competing bacteria, in which the genes controlling production of the ice-forming protein coat have been deleted. Plants are sprayed with these engineered "non-icing" bacteria to prevent the buildup of the normal ice-forming microbes. In California, where fruit crops are especially vulnerable to frost, the technique has protected sprayed crops from frost damage at temperatures as low as -10°C (15°F). As an interesting sideline, the ice-forming bacterial protein is marketed by producers of artificial snow for ski slopes.

Apart from the weather, another major factor affecting crop productivity is soil quality. Modern intensive farming methods usually require large inputs of fertilizer to maintain the level of soil nutrients demanded by high yield crops. But it may be easier and cheaper in the long run to change plants to suit the soil than to change soil to suit the plants. For example, researchers are studying ways to engineer salt tolerance into crops. This could make it possible to expand farmland into marginal areas with poor soils, or even to irrigate fields with seawater.

Another focus of much study is the enormous untapped genetic resources of beneficial soil microbes. One important group of bacteria, *Rhizobium*, lives in

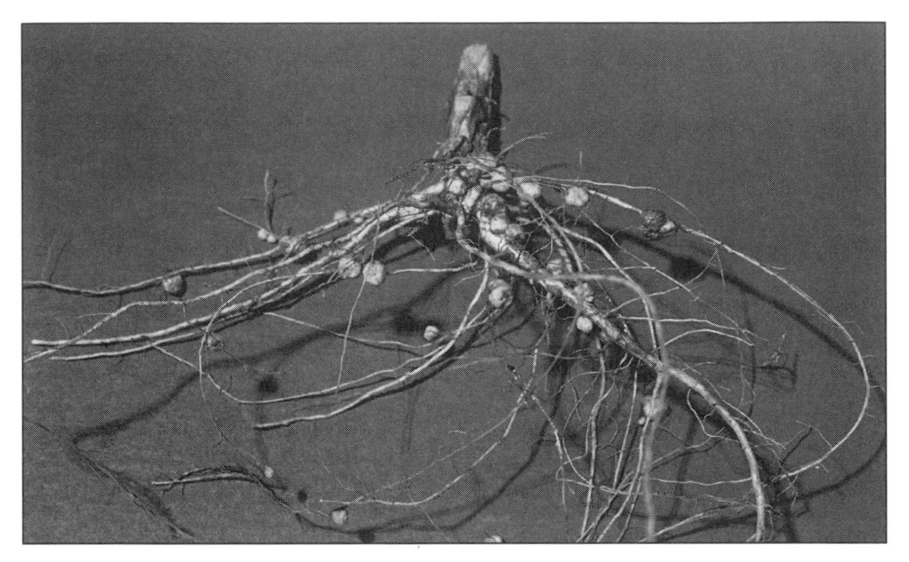

Figure 4.5 Most plants can't survive in nitrogen-poor soils without the addition of nitrogen fertilizers. Not so for legumes like this soybean plant, whose bacteria-filled nodules keep it supplied with a rich source of nitrogen. Can biotechnology be used to help legumes produce even greater amounts of nitrogen in usable forms? Can other plants like corn, wheat, or rice be engineered to have this same "nitrogen-fixing" ability?

nodules on the roots of leguminous plants such as clover and soybeans. These "nitrogen-fixing" bacteria convert nitrogen from the air into ammonia, a form of nitrogen the legumes can use. *Rhizobium* also releases ammonia into the soil where other bacteria convert it to nitrates, acting as natural fertilizers for other plants.

Researchers have identified genes that enhance the nitrogen-fixing process, as well as genes involved in the mechanism by which bacteria attach to leguminous host plants. They hope to modify the genes to boost bacterial efficiency, and to substitute these more efficient strains on the roots of particular crops. It has been estimated that a one percent improvement in the efficiency of nitrogen use could mean a saving of $320 million per year in the amount spent on fertilizers in the United States.

Since root bacteria move freely from the soil into plant root cells, they might also be used as vectors for introducing other genetic characteristics into a crop, such as resistance to root disease. Minimizing root damage could have important spinoff benefits, since healthy roots take up nutrients and water more efficiently, reducing the need for fertilizers and irrigation.

The prizewinning goal in this field is to be able to add genes from these bacteria into non-leguminous crops such as corn and wheat, dramatically reducing the need for nitrate fertilizers. At present, it seems unlikely that this characteristic can be easily transferred. The legume-bacteria relationship is a complex symbiosis, involving many factors that affect the plants' metabolism as a whole.

Farmer-ceuticals?

On tomorrow's farm, the seeds being gathered by a harvester, and the eggs being collected from hens, may not be on their way to the mouths of people or livestock. Genetic engineering is also turning plants and animals into "bioreactors" — living factories for making drugs, industrial chemicals, fuels, plastics, medical products, and other materials. It's an enterprise some call molecular farming.

Plants are already a vast source of natural chemicals and materials such as medicines, solvents, flavorings, fragrances, dyes, oils, wood, fiber, and rubber. By adding a gene here, or taking away a gene there, genetic engineers may be able to change the quality and increase the quantity of these products. For example, altering the structure of fatty acids in oil-bearing plants, such as canola, flax, and soybeans, makes it possible to develop different plant oils, which can be used to manufacture anything from hydraulic fluids to nylon.

Similarly, cornstarch is a versatile and abundant raw material, produced by plants at about the same cost per pound as crude petroleum. By manipulating the genes responsible for starch production, bioengineers can produce starches with different properties. At an Agriculture Canada research center in British Columbia, researchers have developed an edible plastic film from pea starch, pea protein, and canola oil. Products like this might one day be used to package foods such as noodles or soups, allowing the whole package to be dropped into the cooking water and leaving less waste for disposal.

Even more dramatic, transgenic plants can be made to produce completely new products. Engineered canola, for example, is now a source of the blood anticoagulant hirudin, which is normally made by leeches. Hirudin is the most potent clotting inhibitor known, and produces a very low rate of immune reactions in patients. It's a good example of the sort of high-value product that is economically worthwhile for biotechnology companies to develop. Leech genes that code for the small protein (made of only 65 amino acids) are added to canola plants, which then produce hirudin. The hirudin molecules cluster around oil bodies in the plant cells, making it a fairly straightforward process to extract and purify them.

Cows, sheep, pigs, and chickens have been engineered to produce a variety of novel proteins, mostly for use in the medical industry. This approach has several advantages over standard chemical means of production, including relatively low operating costs and unlimited ability to multiply. As an added attraction, bioengineers can ensure that the protein products expressed by added genes are deposited in the milk of mammals or the eggs of hens, making the chemicals easy to harvest and process with little or no detrimental effect on the animals. Here are some of the products already being developed:

- Lysozome is an antibacterial agent that makes up about three to four percent of a normal egg white. Researchers are manipulating the lysozome gene to increase the volume of antibiotic produced and to make lysozome effective against a wider range of bacteria.
- Egg yolk normally contains antibodies that are deposited by the hen to protect the embryo from infection before its own immune system develops. The variety of antibodies can be customized by first immunizing hens with particular antigens. This strategy can now be taken one step further by making transgenic hens. Given genes from other species, these hens will lay eggs with antibodies specific to diseases of, say, pigs, cattle, or people.
- Female mammals regularly produce large quantities of protein in their milk. Scientists can modify the milk content by giving the animals added genes encoding various therapeutic proteins. After the milk is collected, the desired proteins are isolated and purified for use. Products already made in this way, using milk from cows, pigs, and sheep, include human lactoferrin (a good source of iron for babies), antitrypsin (a drug used to treat emphysema), human protein C (needed for proper blood coagulation), collagen (for tissue repair), and fibrinogen (a tissue adhesive).

Closing thoughts

The front page headline in Saskatoon's *The Star Phoenix* read: "Billions in Biotech." It was a fair summary of the focus of North America's first major international agricultural biotechnology conference, held in the Canadian prairie city in June 1996. During the four days of the conference, some 92 speakers from 20 countries spoke about their latest research findings in crop and livestock production, aquaculture, and reforestation. A dominant theme of the meeting, however, and of the local newspaper's coverage, was the opportunity for business expansion.

The gathering was designed to encourage cross-pollination between governments, universities, and industries, with the aim of producing a bumper crop of dollars all round. Saskatoon is the hub of agricultural research in the province of Saskatchewan, and the provincial premier pleased his constituency in his opening address by predicting that sales figures for the region's "agbiotech" companies could jump from $40 million in 1996 to nearly $1 billion by the year 2010.

About two-thirds of the conference speakers delivered lectures on scientific topics such as pest control, livestock reproduction, disease prevention, and soil microbes. The remaining third discussed legal, political, ethical, and financial issues: inseparable traveling companions to biotechnology as it moves throughout the world. The mood of the meeting was optimistic, expressing confidence in agbiotechnology's ability to solve pressing global problems, as well as its promise to create employment and profit.

That conference could not have been further removed

in outlook from a meeting I had gone to only a few weeks before in Victoria, British Columbia. There, about 150 earnest people assembled in a church hall to share their fears about biotechnology. The focus of concern for the speakers in Victoria was also business expansion — and their mistrust of it. The heart of their message was not in specific scientific issues or questions of risk and benefit. Their first question was: do we need biotechnology?

It's a valid question, along with such corollaries as:

- What is the purpose of the technology?
- Is it the best way to solve a given problem?
- Does it improve our quality of life?

These are large questions about human values and welfare, and the desirability of industrial technology. The same questions could be asked about cars and computers, televisions, and hydroelectric dams. As one speaker put it, in the context of criticizing the claims of genetic engineers: "Improvement is not a scientific term."

The contrast between the two meetings made me ponder. In Victoria, agribusiness was demonized as an enterprise interested only in profit. Its opponents clearly saw themselves as defenders of the environment and of human and animal health and welfare. In Saskatoon, the prevailing assumption was that biotechnology is overwhelmingly beneficial. Challengers of this view were seen, at best, as overcautious romantics with little knowledge of science.

It would be a fool's errand to try to reconcile these different views. It is false to suppose that applications of biotechnology must be *either* very beneficial and deserving our support *or* very risky and demanding our opposition. The dispute is essentially ideological, and quickly obscures valuable perspectives under much that is unfair and unproductive. Claims from both sides have a clarity and certainty more at home in the territory of faith than science, and are not to be trusted.

Ideological goal-scoring happens even in scientific debates, when disagreements over the interpretation of data can turn into questions of who is paying for the research. No one, it seems, is immune from bias, not even scientific journals. For example, an issue of *Science* magazine in July 1996 carried the news headline: "Pests overwhelm Bt cotton crop." This headline and the story's opening paragraph painted a portrait of biotechnology's failure, but a closer reading of the details gave a different picture.

Figures given later in the report reveal that no more than one percent of two million acres was actually affected — by just one of three pests the cotton crop was designed to withstand by killing. With conventional insecticides, a loss of about 5 to 10 percent of a crop to pests is routinely accepted. Yet in the atmosphere surrounding biotechnology, a loss of one percent is reported as a failure and a disappointment. So unreasonable are the expectations of the new technology, and so hysterical the reactions when they are not met, that this bit of news actually caused a one-day fall of 18.5 percent in the stock value of the company marketing the seeds.

Beneath the exaggerations and entrenched points of view, however, lies a broad area of common ground. Everyone wants a plentiful supply of nutritious food, economically produced, without harm to the environment. While it is true that companies will only invest in products that make them money, it is also true that they won't make money if farmers and consumers reject their products, or if legislators prohibit them. And while organic farmers and environmentalists have reason to see big business as their enemy, the very same economic and scientific resources that build corporations can also be an ally — for example, in finding alternatives to toxic sprays and artificial fertilizers, and creating niche markets for unusual or traditional products.

Perhaps this perspective is too hopeful, but I was encouraged by a talk I heard in Saskatoon to believe that common sense and pragmatism can prevail. As a university student in the late 1960s, I well remember the impact of Rachel Carson's powerful book *Silent Spring*. An eloquent and farsighted biologist, she was the first to draw wide public attention to the dangers of excessive pesticide use, and was ruthlessly censured by the powerful chemical industries as a result. Corporate scientists denounced her warnings as alarmist and untrue, and questioned her credentials, but in time the consequences of pesticide pollution were too clear to ignore.

Having lived through that battle, I found it gratifying to hear the lecture of a young and enthusiastic chemist working for a giant agricultural chemical corporation in the 1990s. With no sense of irony, he related to his audience the same discoveries that Rachel Carson had been vilified for announcing three decades earlier. Toxic chemicals, explained the chemist, pass along food chains and kill beneficial organisms. They pollute the environment, and promote pest resistance. Having at last accepted these facts, the company for which he worked had developed better strategies of pest control to avoid such results — shorter-lived chemicals with lower toxicity, narrowly targeted and precisely applied. Learning the lessons of ecology, they had even embraced biological control and integrated pest management. Rachel Carson would have been pleased.

Chapter 5
Biotechnology and the Environment

Environmental problems are a combined result of our technology, population growth, and consumer economy. They are problems of resource misuse and overuse. We demand too much, too fast, from the world and produce wastes at every step.

On the grand scale, however, there is no such thing as waste in nature. Discarded molecules are shunted from rocks and soil to plants and animals to air and water and back again, largely through the efforts of microbes. Unseen and unsung, these original recyclers have been neglected until recently by all but academics. With the development of environmental biotechnology in the 1980s, researchers began to look to nature for lessons in how to clean up pollution, monitor environmental health, and produce energy and materials in less destructive ways.

Microbes clean up

Broadly speaking, environmental biotechnology includes any applications that reduce pollution. Methods might use organisms to break down or sequester pollutants (sometimes making useful products on the way), or replace existing activities that pollute with ones that don't. The concept isn't totally new. A traditional example is the septic field, where bacteria are encouraged to decompose domestic sewage so that only harmless breakdown products are released into waterways.

Microbes were first used to treat industrial wastewater as early as the 1930s. More recently, they helped in the cleanup of oil spilled from the Exxon *Valdez* tanker off the coast of Alaska in 1989. Most of the thick oil from the Exxon tanker was initially removed from the pristine Arctic landscape by physical methods — skimming and vacuuming oil from the water surface, and hosing and scrubbing the beaches. But neither scrubbing nor vacuuming nor hosing could remove all the oil trapped between rocks and under

---

Patented oil-eaters

The first patented form of life produced by genetic engineering was a greatly enhanced oil-eating microbe. The patent was registered to Dr. Ananda Chakrabarty of the General Electric Company in 1980 and was initially welcomed as an answer to the world's petroleum pollution problem. But anxieties about releasing "mutant bacteria" soon led the U.S. Congress and the Environmental Protection Agency (EPA) to prohibit the use of genetically engineered microbes outside of sealed laboratories.

The prohibition set back bioremediation for a few years, until scientists developed improved forms of oil-eating bacteria without using genetic engineering. After large-scale field tests in 1988, the EPA reported that bioremediation eliminated both soil- and water-borne oil contamination at about one-fifth the cost of previous methods. Since then, bioremediation has been increasingly used to clean up oil pollution on government sites across the United States.

---

gravel beaches. That's where bacteria were called in, sent after the hidden oil globs like ferrets down rabbit holes.

The bacteria employed in the cleanup use oil as an energy source, breaking down its large, complex molecules into simpler molecules as they do so. To stimulate the growth of these naturally occurring bacteria on the polluted sites, nitrogen- and phosphorus-rich nutrients were applied to the shorelines. The bioremediation project (as this sort of process is called) had some success, increasing the rate of oil breakdown without any lasting harm to the rest of the ecosystem.

By the early 1990s, companies were trying out other microbes on other pollutants. Georgia Gulf and Georgia Pacific corporations tried three types of bacteria on a one-hectare (2.5 acre) site polluted by the toxic compound phenol, and found that the microbes degraded practically all of the pollutant to carbon dioxide and other harmless products in less than 12 weeks. In California, an aquifer contaminated by trichloroethylene (TCE) was treated by first *adding* phenol as an inducer. The phenol stimulated the activity of native microorganisms, which then broke down the original pollutant as well as the phenol.

Microbes in uniforms

Phenol-degrading microbes were first discovered by Dr. Howard Worne, who began research in this field in the early 1950s. He had been commissioned by the U.S. government to develop military uniforms that wouldn't degrade in warm, moist, Asian climates. To everyone's surprise, he found that some microorganisms can break down synthetic fabrics, previously thought to be non-biodegradable. He went on to search for other microbes with the ability to feed on manufactured molecules, and identified the first known organism capable of degrading phenol.

Figure 5.1
Onsite bioremediation
Underground pollution can be cleaned up by injecting microbes and nutrients into the ground.

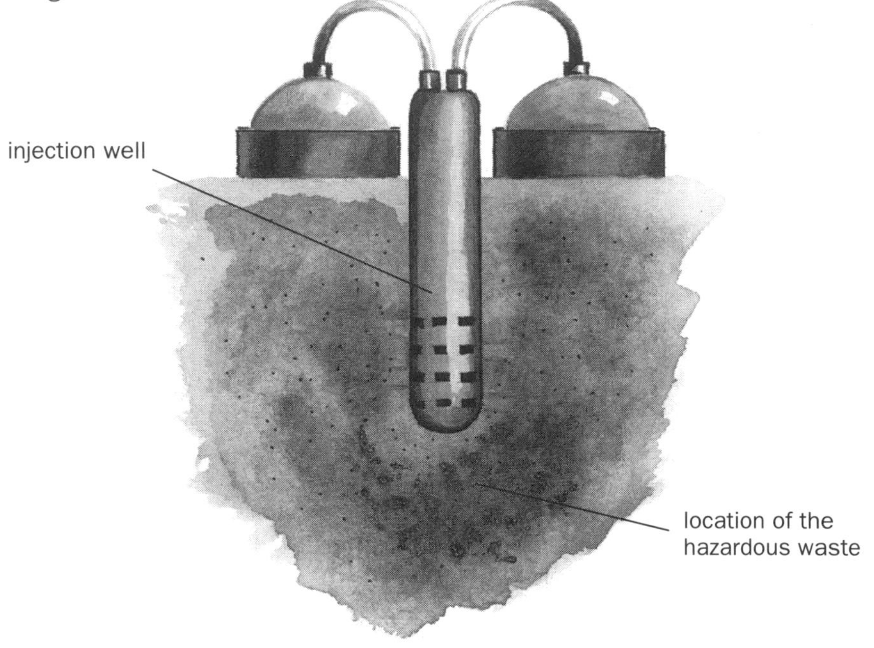

Opportunities for using microorganisms in this way mushroomed as scientists discovered there's practically nothing that can't be viewed as food by one microbe or another. Just as some insects can feed on leaves that are toxic to others, so some microbes can thrive on molecules that would poison most organisms. There are microbes that feed on such toxic materials as methylene chloride, detergents, creosote, pentachlorophenol, sulfur, and polychlorinated biphenyls (PCBs).

The biotreatment of methylene chloride, a suspected carcinogen, is one of the big success stories in this field. Produced by various industrial processes in quantities of millions of pounds each year, methylene chloride can be almost completely eliminated at its source by passing industrial wastewater through bioreactors housing special bacteria. The microbes reduce concentrations of the pollutant from over one million parts per billion to less than five parts per billion — far below the EPA's guidelines. They break down the hazardous methylene chloride into water, carbon dioxide, and salt, thereby eliminating any need to recover it, transfer it, or dispose of it.

Microbes deal with poisonous chemicals in the same way we and other animals process our food, using enzymes to convert one chemical into another, and taking energy or usable matter from the chemical change en route. The chemical conversions usually involve breaking large molecules into several much smaller molecules, much as we break down the complex carbohydrates in our food into simple sugars such as glucose. In some cases, the by-products of a bacterial banquet are not simply harmless but actually useful. Methane, for example, can be derived from a form of bacteria that degrades sulfite liquor, a waste product of paper manufacturing.

Although individual species of bacteria can carry out several different steps of chemical breakdown, most toxic compounds are degraded by groups of bacteria, called consortia. Each species in the group works on a particular stage of the degradation process, and all of them together are needed for complete detoxification.

The search for useful bacteria, fungi, or other organisms to use in bioremediation is best begun on polluted sites themselves. Anything found living there is at least resistant to the deadly chemicals, and may actually use them. After samples are grown and studied in the lab, the most effective strains are shipped back to dump sites and mixed into the chemical brew along with added organic nutrients such as nitrates and phosphates, which help the microbial culture grow. Depending on the nature and extent of the pollution, microbial cleanup can take from a few months to a year or more. When the toxic chemicals are gone, the population of bacteria itself dwindles, being replaced by other microbes native to the area and more suited to the new conditions.

Add liquid and stir

Bioremediation was made simpler and more practical in the late 1980s by a technique that induces colonies of oil-eating bacteria to enter a state of suspended animation — an inactive mode that the microbes normally adopt during extended periods of drought or freezing. In this state, the bacteria can be air-dried, packaged, and stored as a high-concentrate powder with a 90 percent survival rate, to be used in the field as needed. The dried bacteria can be quickly restored to normal function at polluted sites by adding liquid nutrients and biological catalysts.

Within a few years after microbes were first used to clean up hazardous wastes, bioremediation was being described as the most cost-effective means of ridding the earth of its accumulated pollutants. It has already proved successful at many sites contaminated with petroleum products, and is expected to develop into a major industry with sales of more than $500 million by early in the new century. Despite their obvious appeal, however, microbial cleaning crews aren't yet always the first choice of approach.

The main problem with bioremediation is its unpredictability. The effectiveness and speed with which microbes can degrade chemicals at any particular site are affected by many factors: climate, surrounding soil and water, available nutrients, and other chemicals and microbes in the area. Current research is aimed at engineering bacteria to produce more reliable results, as well as expanding the range of bioremediation to include areas contaminated by metals, pesticides, radioactive elements, and mixed wastes.

One of the biggest roadblocks to development is that so little is known about bacterial communities in nature. Although microbes are the most abundant and widespread organisms on earth, their ecology is largely a mystery. The immediate need is to discover how microbial communities function in the wild, and how they respond naturally to stresses, such as exposure to materials that are toxic to most organisms.

Microbes as monitors

Using microbes to carry out tasks in the great outdoors poses a practical difficulty: How on earth do you keep track of what they're doing? It's especially challenging when the microbes are working below the surface of the soil — for example, when they are breaking down underground contaminants. One method of monitoring that has been widely tested involves linking the genes that cause bacteria to degrade contaminants with genes for producing bioluminescence. Bioluminescence is biologically produced light, made, for example, by fireflies, glowworms, some fungi, and many marine organisms. The result of this genetic linking is that the bacteria light up whenever they are working at decontamination.

In experiments already carried out, genes for light production (lux genes) have been coupled to a microbe's genes for naphthalene degradation. The microbe's activity is measured on-site by changes in light level recorded

Figure 5.2 A researcher counts colonies of "reporter" bacteria that fluoresce and are used as a monitor for bacterial cleanup of oil-polluted soils.

through fiber-optic sensors. Being able to study the activity of pollution-fighting bacteria at work in the field in real time is a great advantage, eliminating the need for "best guess" predictions and labor-intensive experiments. It allows scientists to quickly analyze an organism's efficiency and to optimize it, for example by adjusting nutrient levels to boost bacterial growth.

Most current work involving microorganisms in the field uses naturally occurring species, since there is concern over the release of genetically altered microbes into the environment. In controlled field experiments using microbes engineered with lux genes, it is possible to literally watch the spread of engineered genes through the population as the microbes multiply, and to monitor the degree of transfer of these genes from the lab strain to native varieties. This new technology also lets scientists see how microbial activity changes over time as conditions change, and as cultured bacteria become incorporated into a microbial consortium. Finally, the lux genes signal scientists when the cleanup job is done. Where there are no more toxic molecules to find and degrade, the lights go out.

Finding the right microbes for the job

A great deal of research in environmental biotechnology is devoted simply to finding better ways of measuring and sampling the activities of microbes that might be potentially useful. It can take a lot of time and a lot of highly trained microbiologists to find the right microbes needed, for example, to degrade a particular pollutant. Only after the microbes are found can the task of genetic improvement proceed. To speed things up, many biotech companies are racing to develop more cost-effective and less labor-intensive ways of screening large numbers of bacteria. It's potentially a very lucrative field for those that succeed, as screening is the first step in developing most bioremediation programs.

A simple method recently tested quickly sorts out bacteria able to break down volatile organic compounds, such as carbon tetrachloride (used as cleaning fluid), toluene (an anti-knock agent added to gasoline), or xylene. The method consists of growing cultures of different bacteria on plates containing a dye. Eight dozen different cultures are grown on each plate, and the plates are exposed to chemical vapors in a sealed container. Bacteria that can degrade the contaminating fumes carry out oxidative reactions which, in turn, cause the dye to change to a purple color. The precise change in dye color is recorded at intervals by an automated monitoring system, allowing hundreds of cultures to be checked by a single operator. Bacteria initially identified in this way as having good potential for degrading pollutants are then passed on for further tests.

A quick reading

On the slimy underwater surfaces of rocks, stems, and leaves in streams and lakes is a thin layer of microscopic life. This ubiquitous submerged film of algae, bacteria, fungi, and protozoans is called periphyton. It can be used like a book to read the health of its aquatic world.

Monitoring environmental quality is a key task of biotechnology, and what better way to do it than to use organisms themselves as sensitive, built-in record-keepers. The advantage of periphyton as an environmental watchdog is that it is found everywhere and stays put. All scientists needed was to find an easy-to-measure characteristic of periphyton that changes in a consistent and predictable way with pollution levels, signaling changes in the microorganisms' surroundings.

They found such a characteristic in the average lipid (fat) composition of organisms in the periphyton. Lipids are made up of various types of fatty acids. The amount, type, and distribution of fatty acids found in periphyton are clues to the health of the microorganisms. Lipids are used by organisms for two different purposes: as part of their outer membrane structure and as a store of energy. Compared with periphyton from clean waters, samples of periphyton from polluted sites were found to have relatively higher levels of lipids in their membranes. This is probably because the outer surfaces of the microorganisms are damaged by exposure to toxins in the water, and need repair and reinforcement in proportion to the impact of the pollution.

If a sample of periphyton is transferred from an unpolluted stream to a contaminated one, it rapidly changes both its species composition and its fatty acid levels. Given fatty acid readings of periphyton from sites where pollution levels are known, scientists can compare them with samples taken from other sites, and use the differences in this one measure alone to reliably assess the degree of pollution. If the quick readings of the slimy biomonitors show cause for concern, researchers can then go on to measure water quality directly and sample larger organisms.

Microbes and mines

In March 1996, a group of scientists from around the world met in Cornwall, England, to talk about the future of microbes in mining. The subject of their symposium was probably still unknown to most of the public. However, biotechnology was important enough in the metals and minerals industry by then to draw speakers from more than 20 countries, including Canada, Egypt, Japan, Brazil, Ghana, Bulgaria, South Africa, India, Spain, Australia, France, Germany, Chile, and the United States.

Papers read at the conference told how organisms can be used to recover precious metals from tailings and low-grade ores, clean up contaminated soil and water from worked-out mines. Given the range of possibilities and the successes so far, the next few years should see more and more bacteria, fungi, algae, and even plants

Figure 5.3 Floating cattails in mine drainage water provide nutrients for reducing bacteria that live in the root zone and sediments of the water.

being sent to work at the mines. Biomining applications are now firmly on the agenda of mining engineers.

The first success story in the field of biomining involves a group of oxidizing bacteria. One species, *Thiobacillus ferrooxidans*, patronizes copper mines, where it gets energy by oxidizing sulfur-rich ores. For copper mine owners, sulfur is bad news, signifying low-grade ores. Copper in the form of copper sulfide usually needs very high temperatures or corrosive chemicals to extract; both are costly processes. In contrast to these conventional methods, the bacteria release copper from its sulfide bonds as they nibble chunks of ore.

The recovery technique, called bioleaching, involves dumping finely crushed ore outside the mine and spraying it with dilute sulfuric acid to encourage the bacteria to grow. The microbes oxidize the sulfide in the ore and convert the copper to a soluble form. The dissolved copper is washed out of the ore and collected. In this way, the *Thiobacillus ferrooxidans* improves the recovery rates, allowing copper to be extracted economically from low-grade ores and tailings — an increasingly important consideration as much of the high-grade ores have been mined out in the past.

Bioleaching is now routinely used in copper mines in the United States, Canada, Australia, Chile, and South Africa, producing about one-quarter of all copper worldwide. With a value of more than $1 billion annually, bioprocessed copper is one of the most important applications of biotechnology in the mid-1990s.

The potential usefulness of oxidizing bacteria doesn't stop with copper extraction. In Japan, these bacteria are being used to remove iron from mine drainage waters and to treat toxic gases such as hydrogen sulfide. The bacteria oxidize the iron compounds into a form that is much cheaper to manage.

Wherever oxidizing bacteria are used to solve a problem, they also create a pollution problem — they make sulfuric acid. Calcium carbonate added to the acid mine water can neutralize the acid but it forms solid wastes, which must be removed. Acid drainage from abandoned mines is also a serious environmental problem. Here, again, biotechnology is lending a hand in the form of competing bacteria. Reducing bacteria, the counterpart of oxidizing bacteria, can be added to water near old mine sites to counteract the effects of the oxidizing bacteria (see Figure 5.3).

While the acid pollution problems caused by *T. ferrooxidans* are coming under control, biotechnologists face yet another challenge. These bacteria grow slowly, taking about 10 hours to double in number, compared to some efficient bacteria that can double in roughly 20 minutes. Since more rapid growth would be advantageous for large-scale mining applications, biotechnologists have set out to engineer new, quickly dividing strains. But here, too, researchers run headlong into technical difficulties. It is difficult to introduce vectors into *T. ferrooxidans* without first treating them with polyethylene glycol to weaken their cell walls. Afterwards, the bacteria must grow new cell walls. But they are not able to do so when kept in the acid conditions that are optimal for normal growth. It's clear that gene manipulation has not yet become routine.

Hot bacteria

High temperatures are needed for many mining processes, but high temperatures slow down or kill most bacteria. To expand the use of microbes in mines, researchers are studying the genes of the heat-loving bacteria found in hot springs and around oceanic vents. These bacteria thrive in conditions as hot as 100°C (212°F) or higher, and could carry out useful functions in the high-temperature oxidative reactions often used in processing ores.

Golden harvest

There may be gold in them thar hills, but its shine quickly grows dimmer when you add up the cost of labor and equipment needed to extract it, and the pollution created in the process. While Hollywood movie heroes may turn up gold nuggets with only a little light panning, a typical gold mining operation must go through a ton of rock, sand, or gravel to end up with about 1/50th of an ounce of gold. And digging or dredging out the ore is only the beginning.

As well as being one of earth's scarcest metals, gold is also one of its least reactive. One of the few chemicals it readily interacts with is cyanide, which can be a major pollutant in the air, water, and soil around gold mines. Difficult to dispose of cheaply, waste solutions containing cyanide are often kept in open ponds until the chemical is broken down by ultraviolet light.

In the traditional method of processing gold, cyanide is added to crushed ore and the dissolved gold mixture is passed through activated charcoal. Carbon in the charcoal attracts the gold compound from the solution, and the concentrated gold is later washed from the charcoal for final processing.

A difficulty crops up if the ore itself contains naturally occurring carbon — which it does in about 40 percent of the gold mines in the United States. In this case, the dissolved gold compound will bind onto the carbon in the ore and won't subsequently be attracted out of the mixture by charcoal. To prevent this from happening, carbon-bearing ores can be pretreated in a number of ways. Most commonly, the ore is finely ground and roasted at very high temperatures to burn off the carbon. Alternatively, chlorine gas is bubbled through a slurry of ground ore and water to oxidize the carbon. Both processes are costly and polluting, and can make mining operations uneconomical.

A way around the problem is offered by species of bacteria, fungi, and algae that produce and absorb cyanide ions. Adding these microbes to the crushed ore means that further pretreatment is unnecessary, and makes it feasible to process low-grade ores containing as little as 0.02 ounces of gold per ton.

Such methods of using microbes in gold mining have already been patented, but are still being developed and are not yet in use on a large scale. Here's how it happens. Just like feeding time at the zoo, a solution of cyanide-producing microbes is let loose in a holding tank with a slurry of finely pulverized gold ore. As soon as the ore comes in contact with the cyanide produced by the microbes, the gold in the ore oxidizes to form a gold-cyanide complex. The soluble gold is then absorbed by the microbe cells — an automatic chemical process that occurs even if the microbes are dead.

Cells keep working after they're dead

Researchers in the United States have developed a way of using non-living bacteria to decontaminate water polluted by uranium. The bacterial cells, which have an affinity for uranium, are mixed in a polymer to form plastic-like "biobeads." The biobeads are packed into glass tubes and the contaminated water is pumped through them. Any uranium in the water, even in very low concentrations, binds to the biobeads, making the water flowing out of the tubes clean. Technicians are now working on ways to recover the attached uranium and reuse the biobeads.

Biosorption, as the process is called, can be used on carbon-bearing ores without the need for pretreatment. The microbes have an affinity for the dissolved gold that is much greater than that of the native carbon in the ore itself. The microbes easily outcompete the carbon in the gold-binding race.

Many types of microbes produce cyanide and absorb

a gold-cyanide complex, but most do one better than the other. To optimize gold recovery, it's best to use two different microorganisms: one that produces a maximum amount of cyanide and one that is best at absorbing gold. They could be two species of algae, or two different bacteria, or one of each. There are even some species of plants that can do the job, and cultured plant tissue can be used instead of microorganisms. Finding out exactly what works best under what conditions is still being researched.

After the gold has been absorbed by the microbes, the mixture is sent to a settling pond, where the microorganisms sink to the bottom. The separation of microbes from ore can be sped up by adding chemicals, centrifuging, filtering, or screening. The separated microbes are then dried and burned to ash to recover the gold. Each ton of ore yields only about half a pound of dried biomass, and only one to two percent of this is gold.

Bioprocessing gold ore doesn't eliminate the need for cyanide, but it does reduce the amount used and the amount left over as waste. Gold dissolves in even a small amount of cyanide and the microorganisms absorb the gold-cyanide ion complex almost as soon as it is formed.

Using microbes is still a bit of a novelty in the world of mining. Some industry specialists feel there are too many problems to overcome and that organisms are too fragile and limited for use in the harsh metallurgical processing environments common in the industry. Some believe that organisms will prove useful only in limited pretreatment applications, such as leaching, and that biosorption is not currently feasible on an industrial scale. Time will tell.

A new angle to landscaping

It's not only microbes and people that collect gold; plants can also accumulate it from either water or soil. In the early 1900s, in fact, some scientists speculated that plants had played an important part in forming certain gold deposits in rock over geological time. For generations, knowledgeable prospectors have used differences in plant distribution to guide them to buried gold and other metals. But exactly how certain organisms dissolve and concentrate metals has been poorly understood until recently.

The process involves proteins known as metallothioneins. These were first discovered in the early 1980s in horse kidneys, but have since been isolated in nearly every variety of organism tested. Metallothionein molecules have large numbers of atoms that readily bond onto metals such as zinc, copper, lead, nickel, tin, cadmium, bismuth, mercury, silver, and gold. Depending on their particular structure, metallothioneins can be very selective, accumulating one particular metal to the nearly complete exclusion of all others. A metallothionein that selectively concentrates gold was found in 1986 by medical researchers investigating antiarthritic drugs. Some biotechnology companies have been researching the possibilities of synthesizing this protein and using it in gold mining or processing operations.

The power of plants to absorb metals can be used not only to extract precious metals but also to extract what is *not* wanted — metal pollutants in soil and water. An American patent registered in 1994 describes how genetically altered members of the brassica plant family (familiar in such crops as cabbages, mustards, and radishes) can be adapted to absorb toxic or valuable metals through their roots. The plants accumulate metals to levels between 30 and 1,000 times higher than their concentration in the surrounding soil, giving them a metal content of as much as 30 percent of the dry weight of the plants' roots.

Of cabbages and metals

The list of metals absorbed by various members of the cabbage family include antimony, arsenic, barium, beryllium, cadmium, cerium, cesium, chromium, cobalt, copper, gold, indium, both stable and radioactive forms of lead, manganese, mercury, molybdenum, nickel, palladium, plutonium, rubidium, ruthenium, selenium, silver, strontium, technetium, thallium, tin, uranium, vanadium, yttrium, and zinc. The potential for cleanup of polluting metals is tremendous.

Plants used to decontaminate soils must do one or more of the following:

- take up metal from soil particles and/or soil liquid into their roots
- bind the metal into their root tissue, physically and/or chemically
- transport the metal from their roots into growing shoots
- prevent or inhibit the metal from leaching out of the soil.

To be practical, however, the plants must not only accumulate metals but should also grow quickly in a range of different conditions and lend themselves to easy harvesting. The metals, after all, don't disappear when they move from the soil into the plants. If the plants are left to die down, the metals will return to the soil. For complete removal of metals from an area, the plants must be cut and their metal content dealt with elsewhere in a non-polluting way.

Researchers look for suitable properties among both cultivated and wild varieties of plants. Wild species of the brassica family are native to metal-containing soils in scattered areas around the world. The tiny-flowered alyssum plant, for example, is common on serpentine-containing soils in southern Europe. (Serpentine is a magnesium-rich mineral.) Alyssum tend to grow and reproduce slowly, making them unsuitable for large-scale cultivation, but their genes — the ones responsible for metal accumulation — might be transferred into domesticated relatives, which produce several crops per year.

If suitable wild genes aren't available for transplant, researchers can try to get improved metal-storing power in domestic members of the family by inducing more genetic variety. They commonly do this by soaking seeds in a mutation-producing chemical, then screening the germinated seedlings for metal tolerance in artificial solutions containing various metal concentrations. Testing is carried out in batches of at least 50,000 seedlings at a time. The most metal-tolerant and vigorously growing plants are analyzed for their final metal content and the best of them are bred to produce a line of improved plants.

A third option for improving the metal uptake of plants is genetic engineering. This involves introducing genes that code for the specific metallothioneins needed. The genes can be identified and taken from any other species that has them.

Growing metal-absorbing plants can be a cheap, non-polluting, and effective way to remove or stabilize toxic chemicals that might otherwise be leached out of the soil by rain to contaminate nearby watercourses. Or it could be a way to concentrate and harvest valuable metals that are thinly dispersed in the ground.

Fighting chemicals with chemicals

Among the most widespread pollutants in much of the world today are chemicals known as halogenated aromatic compounds. They are commonly found in such products as flame retardants, hydraulic fluids, pharmaceuticals, pesticides, and electrical equipment. Typically, these compounds are chemically inert, water-repelling, toxic — and extremely difficult to get rid of.

An important subgroup of the halogenated aromatics includes pentachlorophenol (PCP), a chlorine-containing chemical commonly used by the wood preserving industry as an ingredient in fungicides, and also found in many pesticides and disinfectants. PCP is highly toxic and thought to cause genetic mutations. Persistent in food chains, low-level PCP contamination is now common in fish, shrimp, oysters, clams, rats, and people. (Tissues of humans in industrialized societies have an average PCP content of 10 to 20 parts per billion.)

Although there are naturally occurring bacteria able to break down PCP in soil and water, they have the disadvantages of living organisms mentioned before: their performance is variable and they need suitable conditions to operate efficiently. So why not find out how the microbes do their work and just use their tools? Japanese scientists developed this approach against another common pollutant, polychlorinated biphenyl (PCB). The scientists patented a technique for extracting the enzymes from a strain of PCB-fighting bacteria and using the extract alone to degrade the pollutant.

To begin figuring out the mechanisms bacteria use to decompose a particular chemical, biochemists start by analyzing the harmless products of their breakdown process, then work backwards to propose which chemical pathways could lead from the original chemical to the products. There might be many possible pathways, each with several steps in it. The key reactions for making PCP less toxic, for example, involve separating chlorine atoms from the large PCP molecules. This might be done by bacterial enzymes cutting the chemical bonds between chlorine and carbon atoms. If scientists can identify and analyze such enzymes, they could then go on to find the genes that encode them. With the genetic codes for the detoxifying enzymes in hand, scientists can insert the genes into new bacterial hosts and clone multiple copies.

It may seem like going around in a circle — starting with bacteria that degrade a pollutant and ending with bacteria that degrade a pollutant. The difference is that the cloned bacteria have an enhanced ability to produce the enzymes involved in the degrading mechanism. In short, they carry out the task more reliably and faster than the original, naturally occurring ones. It's like the difference between a modern dairy cow and its prehistoric wild ancestor.

Making new fuels

Oil is the single most important material fueling our gluttonous resource consumption, but supplies of this 20th-century black gold are expected to run out some time during the 21st century. Oil is also directly and indirectly responsible for a great deal of pollution, not only when burned as fuel, but also when extracted, transported, and refined. Finding a substitute for oil will be a major challenge for the next generation, giving us an unparalleled chance of developing cleaner ways to supply energy to our machinery.

Now what is oil but the product of old dead plant matter? What if we could get something like it out of new plant matter? Ethanol liquid fuel and methane gas are already commonly produced from plants, either from crops grown and harvested for the purpose or from waste plant material, usually from lumber and paper companies or agricultural residues. The appeal is not only that these sources are renewable and relatively clean, but that such fuels can potentially be made in most countries, freeing them from dependency on foreign oil imports.

"Biofuels" are made from plant matter by fermentation. In Brazil, for example, where a warm climate and large land area help make it economical, fuel alcohol has been produced for years from fermented sugar cane. Fuel-making factories are built in areas where the cane is grown, minimizing the need for transport. Cane debris left behind after the fermentable juice is squeezed out is used as boiler fuel, supplying steam for stills, sterilization of equipment, and local electric production. Since no other fuel is required for the operation, the fuel alcohol produced is a net gain.

Not all plant-to-fuel processes are so straightforward or economically worthwhile, however. For example, corn is a common and easy-to-grow crop in cooler climates. It has a high amount of starch, but the starch must be converted into sugar before fermentation can happen. This process takes time and money, and the corn waste has negligible fuel value. Overall, such operations use as much energy for distillation as they get from the ethanol produced. Added to this is the fact that corn is more profitably used as food. This is where biotechnology can step in to help tip the balance. Potential fuel crops can be genetically engineered to grow faster, and with a higher ratio of easily fermentable tissue. Fermenting microbes can be engineered for more efficient conversion of a wider range of materials into fuel, or to alter the fuel products made.

The biggest part of the carbohydrate content in plants is not in the form of either starch or sugars but in cellulose — the material making up the structural cell walls of stems, leaves, hulls, husks, cobs, and the like. Lignocellulose (a mixture of cellulose, hemicellulose, and lignin) in wood and paper waste makes up a vast, cheap, widespread, and largely untapped renewable source of biomass that could be converted to fuel. With the help of microbes feeding on the sugar residues found in waste paper and yard-trash, we could produce huge quantities of ethanol each year from something we now throw away.

However, lignocellulose is very difficult to break down and convert to sugars and then alcohol. The breakdown of cellulose alone releases a mixture of sugars, including glucose, xylose, mannose, galactose, and arabinose. No single organism has been found in nature that can rapidly and efficiently metabolize all of these sugars into ethanol or any other single product of value. The development of microorganisms that can ferment lignocellulose is a major research goal.

An approach to the problem now being taken is to search for the best set of enzymes for carrying out fermentation and then to engineer genes for these enzymes into the most suitable microbes. One useful source of enzymes is the bacterium *Zymomonas mobilis*. Commonly found in plant saps and honey, it has traditionally been used in the tropics to make palm wines and pulque (an alcoholic Mexican drink made from fermented agave sap). The bacterium produces ethanol from sugars at rates far greater than those of common yeasts, and researchers have discovered that it does this using a very short chemical pathway involving only two enzymes.

Attempts made to modify *Z. mobilis* for commercial production of fuel ethanol have met with very limited success, however. The bacterium is particular in its habits and cannot grow if there are no easily fermentable sugars present. Still, if *Z. mobilis* can't do the job, maybe other microbes using its enzymes can. To this end, genes coding for the enzymes have been cloned and transplanted into various bacteria, enabling them to produce higher levels of ethanol from simple sugars.

Speeding up the sugar-to-ethanol conversion is only part of the equation. A harder challenge is to find an easy way to carry out the initial step of converting lignocellulose to simple sugars. Current methods include adding acids and enzymes. The acid approach requires heat and produces acid waste and toxic compounds that can hinder subsequent microbial fermentations. Enzymes are a better alternative and researchers are now trying to modify microbes to produce suitable enzymes.

Electric trees

Fermenting wood and paper waste to liquid fuel isn't the only way to get energy from plant matter. An older and much easier way to convert trees to energy is to burn them and use the heat to make electricity. The Electric Power Research Institute (EPRI) in the United States believes that biomass could make a major contribution to the nation's supply of electricity within the next two decades, using new plantations of genetically altered, rapidly growing crops planted especially for the purpose of supplying energy.

The organization is surveying the country for suitable areas of land with the best soil, nutrients, water, climate, and topography needed to grow and harvest the new crop. Projections of plantation productivity and costs must be compared with current agricultural production in each area to see how they might complement or compete with one another. At the same time, a National Biofuels Roundtable established by the EPRI and the National Audubon Society is studying the environmental impacts and long-term sustainability of producing biomass resources in this way.

Closing thoughts

A cleaner environment, like better health and nutritious food, is something everybody wants. And we want to achieve it cheaply and easily. Environmental biotechnology makes such promises, with its tremendous potential to find better ways to dispose of waste, convert by-products to energy and new materials, and clean up polluted areas. But it sometimes sounds too good to be true. A 1993 statement from the U.S. National Research Council Committee on In Situ Bioremediation said that this field is open to abuse and "has become attractive for snake oil salesmen who claim to be able to solve all types of contamination problems."

With so many variables at play in the environment, it's easy and appealing to say: here's the key that will lead to what you want. It's especially appealing in light of the potentially huge profits that can go to truly successful technologies. While researching this topic, I looked through only a handful of the hundreds of patents for biotechnology "inventions" that companies have rushed to file in recent years in order to stake out their claims in this new territory. It is standard form in patents to claim rights to as big an area as possible, and many companies promise much from little, leaving it to the future to develop ways of applying the scrap of knowledge they reveal.

Take one patent, no worse than others, which announces the discovery of cellulose-chewing amoebas. Amoebas are familiar to most people as the microscopic, jelly-like specks they observed flowing across microscope slides in science classes. Common in both soil and water, most amoebas feed on bacteria, decaying plant and animal matter, or microscopic algae. In the recently filed patent, researchers describe species of marine amoebas that can feed on the cellulose cell walls of seaweeds. They propose that this habit might have several applications in environmental biotechnology.

The amoebas were shown to thrive on the cellulose in seaweed simply by confining them with seaweed as their sole source of food. In addition, the researchers developed a mutant amoeba capable of degrading other large, stable molecules, including polyvinyl chloride (PVC, a type of plastic). These discoveries were presumably patented in the hope that cellulose- and plastic-eating amoebas might one day make money for someone. The patent application speaks glowingly of the microorganisms' potential for reducing problems of plastic accumulation in the environment.

But it's a long way from basic discoveries to applications that are technically practical and economically worthwhile. Preliminary information such as this is often promoted in order to raise funds for further research and development work. As a result, the potential value of a discovery is sometimes confused with its final application. Wishful thinking causes people to talk and act as though something is inevitable, even when there may be many years of work still to be done, and no guarantee at all of success.

Environmental biotechnology will undoubtedly solve some of the problems of pollution in ways far better than any we have today. But it will not solve them all, nor will it help avert the environmental threats of overpopulation or consumerism. In the confusing climate of optimism and fear, greed and despair, that heralds the 21st century, it is well to remember that very few prophecies ever come true.

Chapter 6
Biotechnology in Seas and Trees

The ability of biotechnology to develop new cures, design better crops, and reduce pollution ultimately depends on the properties of living things. It is on our planet's variety of organisms, and their multitude of chemical and genetic resources, that the future of biotechnology rests. Although the applications of biotechnology today use mainly familiar organisms from labs and farms, there is a vast untapped well of life around the world from which tomorrow's successes may flow.

Serendipitous discoveries from unlikely sources, such as the anticancer drug found in Pacific yew trees, hint at the possibilities to come. However, the sad truth is that we know next to nothing about the vast majority of living things that we share the planet with. All but six of the world's 33 major groups of animals are mainly marine, yet we know less about the oceans that cover two-thirds of the globe than we do about the moon. Tropical and temperate forests are home to a greater abundance of life than any other land ecosystem, yet we are clearing land of living forests at an unprecedented rate. Seeing no more to trees than fuel and building materials, we've squandered unknown numbers of organisms with every vanished hectare, not even knowing what we've lost.

An ocean of opportunity

A chart of evolution on my wall shows the major groups of animals rising and branching from a submarine stem of single-celled ancestors, growing upward through a deep blue sea to thrust their topmost twigs above the surface and into the air. It's a vivid reminder that not only did life begin in the water but that the great majority of organisms today remain aquatic. Only reptiles, birds, mammals, insects, and plants have truly conquered dry land. By far the bigger share of earth's living creatures — fish, urchins, crustaceans, worms, mollusks, anemones, sponges, microorganisms, and the rest — still have all or most of their branches underwater, bearing a vast submerged wealth of genetic information.

It's ironic in this century of scientific exploration that we've spent so much effort and imagination looking for evidence of life in outer space while almost ignoring the many small, strange beings we know live here with us, below the waves. Adapted to some of the planet's most extreme environments, some of them thrive in sci-fi-like conditions: the intense pressure, freezing cold, and perpetual darkness of the ocean depths; boiling mineral streams gushing from sea floor vents; and the punishing regime of pounding waves and drying winds along shorelines. These marine organisms are unique repertories of strategies for survival, and the unfamiliar tools they use to meet the challenges of ocean life could turn out to be invaluable resources.

The deep ocean is a much harder place to explore than outer space, but spinoffs from space technology have given us new ways to uncover submarine mysteries.

Crewed submersibles, remotely operated vehicles, geosynchronous satellites, sophisticated acoustic measuring devices, pressure-retaining deep-sea samplers, and computerized databases, for example, have helped reveal bizarre worlds of unexpected life on the ocean floor.

In the 1980s, giant tube worms were recovered next to deep-sea hot vents, and new species of mussels were found around methane seeps in the Gulf of Mexico, with symbiotic methane-using bacteria living on their gill tissues. More new animals, plants, and microbes are discovered regularly, while a growing number of studies show the potential value of already-familiar marine organisms, many of which make substances unlike any found on land.

Every new discovery brings the chance of finding useful materials and techniques. For example, a better understanding of how shellfish form their shells has helped scientists create fine ceramic coatings, an application currently being introduced in manufacturing auto engines and medical equipment. A single novel finding can go a very long way: one compound extracted from a Pacific sponge has already spun off more than 300 chemical analogs (similar compounds), many of which are being tested as anti-inflammatory agents.

The marine environment is a particularly fertile source for new bacteria, especially since, according to some estimates, less than one percent of the earth's bacteria have been isolated and described. Previously unknown forms of ancient cold-water bacteria were recently discovered at depths of 500 m (about 550 yd), yet despite their abundance and importance in marine ecosystems, practically nothing is known about them, or countless other marine bacteria. New technology has revealed that viruses, too, are abundant in seawater samples. Some of these viruses infect species of marine phytoplankton (small, drifting plants) and could have a major effect on the vital process of photosynthesis in the oceans.

Submarine explorers are also still being surprised by larger forms of life emerging from the deep. In the early 1990s, a research student at the Scripps Institute of Oceanography in California encountered the densest aggregations of animal life ever found on earth. Living in large mats of plant debris in an underwater canyon off the coast of southern California are billions of tiny marine crustaceans. Up to three million of the minute crab-like creatures crowd together in 1 m^2 (10 sq ft). Scuba divers collecting samples at the canyon regularly saw large numbers of fish feeding at the mats. If the huge amount of living matter in this hidden canyon is duplicated along other coastlines, it could play an important role in ocean food webs.

At the same time as these exciting new discoveries about sea life are being made, we risk losing forever what we have barely just found. Overfishing and the pollution of coastal waters threaten the very survival of many marine ecosystems, a potential catastrophe brought about by our own indifference.

Foul is fair

Drifting along on the open sea might be a romantic's dream, but for many small marine animals, drifting is something to avoid at all costs. Going with the flow could take them thousands of miles from their origin, exposing them to new risks or making it harder to find mates. Much better to stick where they are and resist the current's pull. Sticking is what a lot of sea creatures do extremely well, as anyone who has had to scrub a boat hull well knows. Hard, smooth surfaces are at a premium in the sea, and it doesn't take long before submerged structures are covered with a slimy coat of bacteria and algae.

Fouling, as this growth is called, not only increases drag on moving vessels but clogs industrial pipelines and speeds corrosion on metal surfaces. Pioneering colonies of microbes pave the way for later settlement by invertebrate larvae and seaweed spores, building up to "hard fouling" by barnacles, muscles, anemones, and other organisms that eventually demand costly removal.

It's not only undersea where aquatic microbes like to grow on reefs and sunken ships. Microbes can attach to any exposed site in contact with watery fluids, causing problems in places such as heat exchangers, trickling filters, or aquaculture circulation systems, and even in human medical implants and prosthetic devices. Since it's not possible to use toxic chemicals to deter squatters from these places, nor easy to remove them by scrubbing, scientists must look for more ingenious ways to avoid fouling.

New answers might come from studies of how marine organisms attach themselves, and of how large aquatic animals and plants prevent their surfaces from being settled on. Researchers have already investigated the genetic coding for biological adhesives used by other organisms, such as nitrogen-fixing bacteria that glue themselves to the root nodules of leguminous crops. Disease-causing bacteria that adhere to mucous membranes in the nose, throat, and lungs also use biological adhesives. Knowing which genes allow marine bacteria to produce their "glue," and analyzing which cues in the environment regulate the genes' expression, it may be possible to develop a means to turn off the genetic switch and keep underwater surfaces free of settling microbes.

Marine organisms that settle down to a sedentary life must not only find a parking spot and stick to it, they must also be able to protect themselves from other creatures that want to settle on or near them. To do this, many attached animals produce defensive chemicals, which they release into the water to create a protective zone around them. The chemicals inhibit larvae or microorganisms from settling, as well as deter predators.

Underwater plants also produce repelling compounds to prevent bacteria from attaching to them, and some have surface structures that neutralize bacterial adhesives. For example, a chemical made by a species of eelgrass effectively prevents its leaves from being fouled by bacteria, algal spores, barnacles, and tube worms. If biotechnologists can analyze the chemical and find a way to manufacture it in quantity, they would have a novel defense against fouling.

Underwater drugs

Drugs — both legal and illegal — are simply chemicals that affect how living things function, by interacting with particular parts of particular cells to change the way they work. They are no different in principle from some of the chemicals our own bodies produce, such as hormones and enzymes. Every living thing makes its own set of drug-like chemicals for its own purposes, and the only sources of drugs before the establishment of large-scale drug manufacturing in the 19th century were plants, animals, minerals, bacteria, and fungi. While the thousands of different pills, powders, and liquids dispensed by pharmacists today are nearly all synthetic purified chemicals, each designed for a very specific and limited purpose, the roots of the drug industry remain in naturally occurring chemicals.

Pharmaceutical researchers continue to analyze plants and microbes for potential new drugs, but have barely begun to study marine sources, especially bacteria, fungi, and algae. According to W. Fenical and P. R. Jensen's review of the subject in *Marine Biotechnology*, "On the basis of the few chemical studies reported, and in recognition of the unique compounds that have been isolated, it can be concluded that marine microorganisms could, if effectively explored, represent a major biomedical resource."

It's been discovered, for example, that some of the brightly colored green and purple sulfur bacteria that live symbiotically with sponges and sea squirts produce potent chemicals that can stop viruses in their tracks. A new compound taken from deep-sea bacteria living in sediments more than 300 m (330 yd) below the sea surface inhibits the replication of HIV, the virus that causes AIDS. And still other marine microorganisms are hot tips for new antibiotics, much needed against strains of disease-causing microbes that have developed resistance to known drugs.

Anticancer agents are high on the list of substances being sought by researchers at the Scripps Institute of Oceanography. One of their discoveries is a marine plant adapted for growing in almost saturated brine. It produces a variety of chemicals, including beta-carotene, a possible anticancer agent. Some marine plants in the

Antarctic Ocean, which receives nearly six months of continuous sunlight, make compounds that absorb ultraviolet radiation, a feature with potential to protect people from skin cancers.

Many sponges and corals make chemicals that have been used successfully to treat the inflammation and pain of acute asthma, arthritis, and injuries. In most cases, these marine products don't have the problems and side effects of steroids, Aspirin, and other conventional anti-inflammatory drugs.

Sharks, which in the wild quickly recover from terrible injuries and rarely get ill, are an especially rich source of potential medical treatments. Shark blood contains antibodies against a huge array of bacteria, viruses, and many chemicals. More remarkable, sharks appear to be completely immune to cancer. Even when injected with potent cancer-causing chemicals, which induce the disease in every other type of animal tested, sharks remain cancer-free. One idea being tested is that the sharks' cancer resistance is due to proteins in their cartilage. Materials made from shark cartilage are also being used to make artificial skin to protect burn victims against infection.

A cornucopia of chemicals

Forty seven billion dollars' worth of pesticides are sold each year in the United States alone, and the search for new and more effective chemicals to control pests is neverending. As with germs and antibiotics, the living targets of pesticides eventually evolve resistance to particular toxins, and their populations begin to climb again. To keep control, pesticides themselves have evolved over the years, from the synthetic, widely toxic sprays and powders of the 1950s and '60s to chemicals with more specific targets and less harmful impact on non-targeted organisms. Biopesticides — chemicals derived from animal and plant sources — still make up less than 10 percent of the huge pesticide market, but new products from marine sources are expected to add to their arsenal in the next few years.

One marine biopesticide being used today is Padan, which was developed from a bait worm's toxin known to Japanese fishermen for centuries. Padan acts against the larvae of such pests as the rice stem borer, the rice plant skipper, and the citrus leaf miner. Other insecticides, effective against grasshoppers and tobacco hornworms, are based on chemicals produced by sponges and sea slugs to deter feeding by fish. These toxic chemicals include terpenes, a broad class of compounds also used in solvents and perfumes.

Some industries and researchers are particularly interested in marine organisms that thrive in extremely hot sites, such as around hot vents on the sea floor. Even moderate heat causes most biologically active molecules to give up the ghost and stop working, so it's especially valuable to understand how hot-water microbes can still grow at temperatures over 100°C (212°F).

Enzymes that function at high temperatures are called thermostable enzymes. Those that modify DNA molecules — for example, polymerases, ligases, and restriction endonucleases (explained in Chapter 2) — have proved invaluable in studies of genetic material. A thermostable DNA polymerase enzyme produced by bacteria living in hot springs in Yellowstone National

Park was the basis for the first heat-cycled polymerase chain reaction (see page 51). This enzyme was named Molecule of the Year by *Science* magazine in 1989. A second generation of thermostable PCR enzymes has already been harvested from bacteria living near thermal vents on the ocean floor, and are marketed by a small company in Massachusetts as Vent® and Deep Vent® Polymerases.

Other marine biochemicals that interest industry include:

- salt-resistant, protein-digesting enzymes secreted by marine bacteria, a potential ingredient in detergents used for cleaning some industrial equipment
- compounds made by algae and sponges that promote germination and growth of plant roots and leaves
- unique enzymes that help combine biological molecules with halogens (such as chlorine and iodine). Japanese researchers are extracting these in large amounts from marine algae to use in the medical, cosmetic, and food processing industries.

Car parts from crabs

A company in Chicago has developed a new class of biodegradable materials after studying how marine mollusks and crustaceans produce their light but sturdy shells. A key aspect of these elaborate mineralized structures is the fine scale on which they're built — a microscopic scale measured in nanometers. (That's one-billionth of a meter.) Nanometer-scale structures have unusual and useful properties, which engineers hope to understand and use to create novel bioceramics for use in such products as medical implants, electronic devices, protective coatings, and automotive parts including fuel pumps.

Another aspect of shell development that researchers are copying is the way in which shells are deposited in thin layers from mineral solutions at low temperatures. Two companies in the United States are developing a technique to mimic this process as a method to line plastic fuel pump components with a ceramic skin, so they can be used with alcohol-based fuels. A medical technology company plans to coat bone replacement parts with calcium phosphate in the same way, producing prosthetics with improved compatibility with body tissues and increasing the rate of new natural bone formation.

Research and development

While investigators strive to apply the properties of marine molecules, a huge amount of research is aimed at finding those useful molecules in the first place. New methods of analyzing molecular properties on a large scale have stimulated the booming biotechnology industry as much as actual product development itself.

In the not-too-distant past, testing new chemicals for their biological effects was a long, laborious, and labor-intensive process. It required multiple, separate measurements, and large samples of the substances being analyzed. Today, the task is made much faster and simpler by automated assay techniques needing only minute amounts of chemicals to work with. These sophisticated tools make it possible to screen hundreds of newly discovered compounds in a short space of time, testing each for a wide range of biological activities.

The problem with collecting potentially valuable chemicals from sea animals is that living things don't always turn out their secretions predictably and on schedule like factories. Many of the most interesting natural substances are produced in limited quantities at restricted times — for example, only in response to stress, or at certain stages in the life cycle, or in certain seasons. Production can also be influenced by an organism's nutrition, its location (including depth), physical and chemical conditions of the water, or by other organisms around it, such as predators, parasites, or other members of its own species.

The same things make it difficult to produce large amounts of particular chemicals from captive organisms, or to stimulate their production in cells or tissues taken from plants and animals. A more effective approach would be to transfer the necessary genes into microorganisms that can be easily cloned and cultured. But not all the biochemicals found in an organism are the direct products of its genes, produced by it for some purpose. Many are indirect by-products — generally wastes as far as the organism is concerned. To manufacture those, it may be necessary in the first place to identify the enzymes used in the chemical reactions that produce the by-products.

Cholera kits

Until recently, there was no quick or easy way to test for the disease-causing cholera bacterium, which lives in coastal waters where clams, oysters, and other shellfish grow. Now scientists have developed and marketed a simple hand-held kit that can detect the microbe in samples of water or shellfish tissue in a matter of minutes. The assay is based on an enzyme-linked immune reaction and produces a visible color if the cholera bacterium is present. Kits were sent to South American countries in 1992 to help health officials combat a cholera epidemic.

Fuels from the sea

In the last chapter, I wrote about making fuel from biomass. The principal process that produces any kind of biomass is photosynthesis, a series of reactions that convert carbon dioxide and water into carbohydrates. Since there's 50 times as much carbon dioxide in the oceans as in the atmosphere, not to mention plenty of water, the oceans are potentially a rich source of biomass production. (Marine plants turn more than 30 billion tonnes of carbon into biomass each year.) The problem with marine biomass, however, is that it isn't easily harvested and, in any case, biomass isn't competitive with other cheap and abundant types of fuels. But what if the process of photosynthesis itself could be modified through biotechnology?

The natural plant enzyme that captures carbon dioxide for photosynthesis, for example, is relatively inefficient. If computers could redesign it for optimal function, and scientists could engineer genes to produce the remodeled enzyme, they could develop a new breed of photosynthetic marine organisms to make more biomass more efficiently. As well, the chemical composition of the biomass they grow could be altered, customizing it for particular uses. For example, algae with a higher fat content than standard-issue plants would be a better source of fuel.

Another way of increasing ocean productivity is to modify the nutrient balance of the waters. Plants use nutrients such as carbon, sulfur, phosphorus, and iron for important metabolic processes. A shortage in any given vital element could be the weak link in the chain that sets the limit on their biomass production. Scientists are using biotechnology to understand these cellular processes better and to pinpoint crucial biochemical pathways. For example, scientists were able to create a significant increase in plant production by adding an iron-containing solution to open waters off the Galapagos Islands.

Farming the seas

One of the most talked-about uses of the sea is for raising fish, mollusks, crustaceans, algae, and other edible marine organisms in captivity or semi-captivity. Readers who have heard for years about the importance of fish farming to feed a growing human population can be excused for raising skeptical eyebrows on hearing of it yet again. Although specialized operations for growing seafood have been established in many coastal regions around the world, the industry itself hasn't grown as fast as predicted. Logistical and economic factors combined with pollution and diseases have weighed against the success of many fish farms. But with biotechnology added to the scales, the balance may be tipping the other way.

Asian countries are far in the lead of marine farmers, currently producing 85 percent of the world's annual supply of 21 million tonnes of cultured aquatic organisms. Other fishing nations are striving to catch up, with the Canadian government, for example, forecasting that aquaculture production will contribute at least a quarter of the total landed value of the nation's fish harvest by the end of the 1990s — a big rise over its 1987 level of only three percent.

In the United States, the aquaculture industry grew during the 1980s with the success of catfish farming, but the cultivation of marine species still languishes. Despite its long coastline, the United States ranks 10th in the world in the value of its aquaculture products. Imports of seafood and seafood products contributed $2.5 billion to the American deficit in 1992, behind only petroleum, automobiles, and electronics. Add to such national shortfalls the decline in major fisheries worldwide, and the need to develop cultivated stocks is more urgent than ever.

The aim of aquaculture producers is the same as that of any farmers: They want to produce bigger, healthier animals and plants in the quickest, most efficient way possible. Like farmers, their tools include better feed and health care, growth hormones, reproductive technologies, and genetic engineering.

Biotechnology can help produce more seafood at many different points in the fish-farming process. It can be used to speed an organism's rate of growth, lower its age of maturity, increase egg production and fry survival, and develop a year-round capacity for reproduction. Genetic engineering could improve disease resistance, the efficiency of converting food into flesh, and the quality and composition of that flesh.

Among the first of the new techniques used to boost farm fish production during the mid-1980s were synthetic growth hormones. Hormone-fed fish in those early trials gained weight up to twice as fast as normal fish, but their rapid growth didn't mean a growth in profits for fish farmers. The manufactured hormone was expensive, and a proportion of it went through the fish without being absorbed. At the same time, public reaction against the use of growth hormones in the dairy industry didn't encourage fish farmers to believe there would be ready acceptance for their hormone-fed fish.

The next approach was to inject genes for producing natural growth hormones into fish eggs. While this genetic engineering technique isn't always successful, it takes only a few hundred altered eggs to produce a viable breedstock that can perpetuate the gene in its lineage

Figure 6.1 There's a huge difference in size between genetically engineered (above) and regular coho salmon (below), both 14 months old.

through normal reproduction. The gene is there to stay, passing along its benefits at no extra cost to the fish farmer. However the effect is achieved, fish with added growth hormones not only grow many times faster than normal, they convert food into flesh with as much as 15 percent greater efficiency.

With stocks of fast-growing fish established, breeders can further raise their production by increasing the fish's breeding output. Once again, hormones are the key to this change. An early method of inducing spawning was simply to inject fish with sex hormones, but this technique requires repeated injections, which are stressful to the fish and time consuming for the operator. To improve on this, scientists developed a single-injection technique in which the hormones are embedded in a matrix of large molecules; the hormones are released slowly from the matrix over a long period of time leading up to spawning. This method is now used in farming striped bass and other fish.

In some operations, ironically, fish farmers risk losing much of their marketable stock because their fish mature too soon. If fish come into breeding condition at too young an age, they increase operating costs and may be too small to sell profitably. Precocious maturation of salmon was a big problem for salmon farmers in British Columbia until technologies for controlling reproduction turned things around. Using both hormonal and genetic controls, researchers developed techniques for sterilizing fish and developing strains in which few or no males were produced. Approximately 80 percent of the chinook salmon currently cultured in British Columbia are all-female strains. Sterilizing those grown for market prevents them from maturing, giving operators better control over marketing time and product size.

Diseases and pollution

The dense populations, genetic uniformity, and stress found on commercial fish farms make them a paradise for bacteria and viruses. Where large commercial aquaculture operations grow and spread, so, too, do many infectious diseases. The threat of disease may, in fact, be the biggest factor limiting the development of aquaculture worldwide. Many millions of dollars are at risk in countries like Japan and Taiwan, where intensive aquaculture systems can produce over 2,000 kg/ha of shrimp (11,000 lb/acre). Many governments restrict fish trade because of disease problems. In the United States, more than 50 diseases affect cultured fish and shellfish, causing losses of tens of millions of dollars annually. And the risk may not be only to fish. A type of meningitis caused by bacteria in fish was recently reported to have infected a number of people cleaning tilapia raised on fish farms.

With such a huge demand for disease control, more veterinary medicine companies are using biotechnology to find new ways of growing healthy fish, based on natural differences between resistant and susceptible animals. Vaccination is a key area of disease prevention, aimed at producing immune fish with less need for antibiotics and other drugs. Antiviral vaccines, for example, have been mass-produced by cloning parts of the viral protein coat in bacterial cultures.

The health of the booming aquaculture industry itself is vitally dependent on the quality of the environment. About 80 percent of marine pollution originates on land, with the outpouring of sewage, pesticides, heavy metals, radioactive wastes, oil, sediments, and other materials into streams and rivers. Because most of the fish we eat are predators near the top of food chains, they tend to concentrate many of these pollutants in their bodies, creating a health risk for human consumers. Shellfish, too, concentrate toxins in the surrounding water by their method of filter feeding.

While fish farms are susceptible to pollution, they create pollution and environmental degradation of their own. Shrimp farms in Asia have created salination problems in surrounding land and water, and caused the loss of large areas of natural wetlands. The escape of genetically modified species and the release of medical drugs into the environment raise the concern that wild populations of fish might be endangered by fish farming. The loss of wild fish would mean the loss of potentially valuable genes as well as reduction of biodiversity.

Changes in wild species can now be closely monitored using the tools developed for analyzing and identifying genes. This allows scientists to define species, stocks, and populations that may appear similar to the eye but have important differences in their gene pools. Improving techniques for breeding captive animals, and technologies for preserving frozen eggs, sperm, and embryos, can also help conservationists restock depleted areas and maintain threatened species.

Frozen fish

Since North Atlantic fish stocks fell so dramatically in the 1990s, many fishing communities scattered along Canada's eastern shoreline turned to aquaculture of salmon and other fish. But the more northerly communities face the challenge of protecting their captive fish stocks (especially young ones) from the cold. During the Canadian winters, much of the east coast has sub-zero seawater temperatures. These conditions would freeze halibut and Atlantic salmon raised on fish farms, making the use of sea cages in these areas all but impossible — unless stocks of freeze-resistant fish can be developed. And that is exactly what researchers at the Memorial University of Newfoundland are aiming to do, with good results so far.

The researchers have been experimenting with an antifreeze gene found in a species of Arctic-dwelling flounders. The gene protects the flounder's body fluids from freezing by instructing its liver to secrete proteins that inhibit ice crystals from forming in the fish's blood. The gene from the wild fish has already been cloned and inserted into Atlantic salmon, producing a stable line of transgenic fish that can withstand frigid water temperatures much better than their farmed but unengineered relatives.

The forest and the trees

Biotechnology helps us look at forests with fresh eyes. In their traditional role, forests provide wood for planks, pulp, and firewood. Biotechnology makes it possible to develop faster-growing, disease-resistant trees, increasing the production of these renewable materials. In a second, more valuable role, forests are reservoirs of biodiversity. Their potential genetic resources could be vital to the development of such human needs as improved drugs, pesticides, foods, and materials.

A harvest of wood

To sustain a tree harvest into the future, forest managers need to replace what they cut — an obvious if late-acknowledged truth. Where forests have grown undisturbed for centuries, they are impossible to replace, but the next best thing is to substitute tree plantations, in which the focus is on planting improved varieties of trees.

Trees are a crop that can outlive their growers, and people in past generations planted trees that their children would harvest. With such a long growth period, it was hard to select seeds to develop improvements from generation to generation. Now there's no need to wait so long. Biotechnology speeds up crop rotation time and gives better selection methods, letting tree growers compare genetic varieties in a matter of years rather than decades. Genetic engineering and mass cloning are radically changing the rate and efficiency at which tree improvement can be achieved.

The traditional method of reforestation, especially in coniferous forests, is to collect seeds from the most desirable trees, germinate them, and plant the seedlings. Superior seed-bearing trees are selected and grown in seed orchards for ongoing production of genetically improved seeds. The problem with letting conifers make seeds on their own, however, is that you don't have control over who the father is. While you can select the best mother trees for your orchard, the father's pollen, which contributes 50 percent of genetic information in the next harvest of seeds, can stray in on the breeze from any old tree nearby.

Controlled pollination, in which the maternal tree is guarded from all but a superior tree's pollen, can alleviate this problem but adds to the cost. And even when both parents are carefully selected, they give a crop of sibling seeds with many different genetic combinations. Since not all the combinations will be favorable, the potential genetic gain is reduced. A better alternative is to switch from sexual reproduction to asexual reproduction.

Asexual reproduction means, essentially, cloning offspring from a single parent, with the advantage that all the qualities of the parent are known and will be passed on without dilution. Grafting, vegetative propagation, and micropropagation are all techniques used for this purpose.

Grafting is commonly practiced in horticulture by taking growing shoots of the desired plant and connecting them to root stocks of closely related plants. It's a technique that requires time and skill and isn't practical for large-scale reforestation.

Vegetative propagation, better known as taking cuttings, involves cutting growing stems and getting them to root. Although many conifers can be propagated by rooted cuttings, large-scale production is again extremely costly due to difficulties in automating and mechanizing the process. Also, it is more difficult to root older plants, limiting the potential of propagating valuable trees to those in their first few years of life.

Micropropagation is the most recently developed method of cloning, and the one with the greatest prospects in the forest industry. This method of production has three main advantages:

1. It can be easily automated and mechanized to turn out the large volumes of planting stock needed for reforestation.
2. Plant cells can be preserved almost indefinitely in liquid nitrogen, giving growers access to valuable genes far into the future.
3. Cell cultures can be genetically engineered and cloned to produce stocks of transgenic trees.

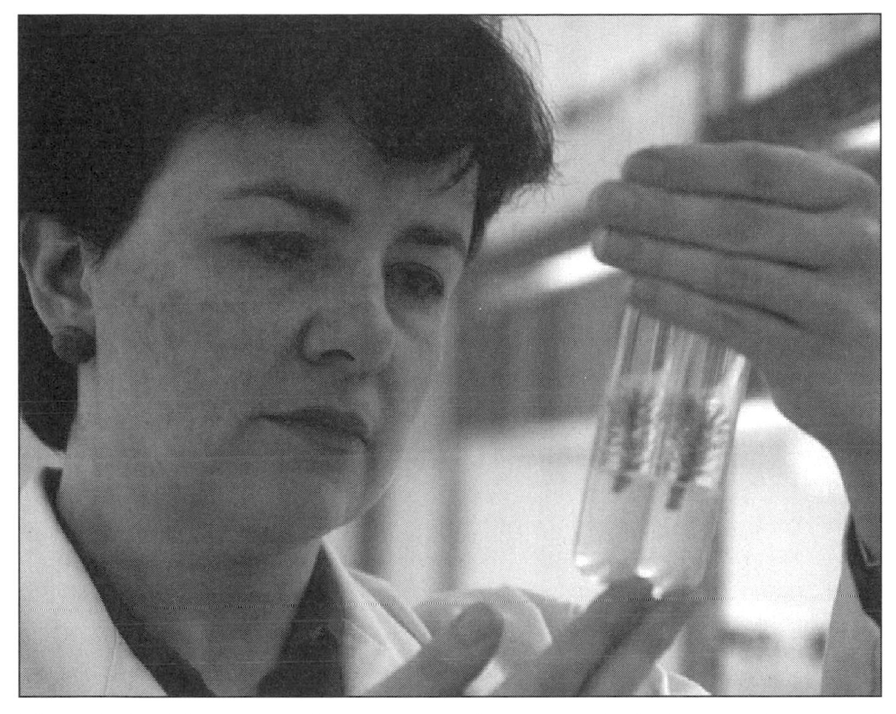

Figure 6.2 Micropropagation means producing new plants from small pieces of plant tissue or individual cells.

To date, experiments have shown that some species of coniferous trees are much easier to micropropagate than others, and investigators are trying to find out why. The two most economically important groups of conifers growing in the northern hemisphere are spruces (about 30 species) and pines (about 95 species). Efforts to micropropagate spruces have been consistently successful, while attempts to reproduce several species of pines in this way have been striking failures.

The best source of material for micropropagation is seed tissue, taken from fertilized seeds containing developing plant embryos. The embryos are picked apart and each separate piece is stimulated to grow into a new, identical plant. The growing microscopic clones are called somatic embryos — embryos derived from the body (soma) tissue rather than from the reproductive cells.

Among spruces, success with growing somatic embryos is often as high as 95 percent when starting with immature embryo tissue, and as much as 55 percent even when tissue taken from fully developed, dry seeds is used. Many of the somatic embryos initially grown from a few cells in a lab dish go on to become established seedlings in a nursery. Among pine species, however, the results are dramatically different. Researchers have been unable to get more than about five percent of their pine cell cultures to develop into somatic embryos, and have had almost no success trying to coax the embryos to grow into small seedlings.

Trying to select and breed pine stock with better embryo-producing potential runs the risk of eliminating genetic varieties that, while poor at producing somatic embryos, might be good for other desirable qualities such as rapid growth. Researchers continue to look for ways around this problem. One question they've focused on, for example, is how changes in the nutrient medium and other environmental conditions in the lab might improve somatic embryo development in pines.

In the meantime, micropropagation of spruces races ahead. Production systems have been developed for bulk-handling tissue cultures and semiautomated planting of germinated somatic embryos. The main emphasis of commercial producers is to provide genetically uniform stock with insect resistance and increased growth rate. One company in Canada shipped 150,000 spruce somatic seedlings to nurseries in 1995, and is developing other varieties of spruce for the ornamental conifer market.

Building better trees

As techniques for mass-cloning trees from tissues become better established, scientists are applying them to more different species, including hardwoods. In the United States, researchers have used micropropagation on sweetgum, alder, and silver maple. In addition, tissue cultures have been developed as a step towards genetically altering black locust trees.

New genetic characteristics can be introduced into a breeding stock of trees either by transferring genes into somatic embryos from another organism, or by testing genetic variability in different tissue cultures and selectively cloning those with the desired qualities. Scientists with the U.S. Forest Service in Wisconsin are using both these techniques to develop herbicide resistance in hybrid poplars. Researchers in Minnesota are inserting genes for insect resistance in somatic embryos of black locust.

A quick way of finding out whether particular plants have any of the desirable genes wanted for crossbreeding is to use gene probes. This technique can be used to monitor the genetic makeup of hybrid seeds produced in orchards and track the degree of inbreeding. Such long-term research on gene expression and genetic transformation is making genetic engineering and cloning almost routine in some parts of the forest industry.

Pests and diseases have always been a major threat to tree nurseries and reforestation sites, and biotechnology has added tools such as genetic modification and vaccination to the traditional weapons of chemical sprays. Spruce trees genetically engineered to resist spruce budworm infestation were first developed in 1993. Another novel approach to protect trees from pest damage is being studied in Finland, where researchers are breeding types of birch unpalatable to moose, hares, and voles. But, as in agriculture, it would be a mistake to focus only on genetic and therapeutic remedies and forget the importance of good cultural practices when growing trees. A healthy soil is vital, especially for species of conifers that rely on symbiotic relationships with soil fungi to help them obtain nutrients.

A key part of the growing environment that could have a long-term impact on all reforestation projects is the climate. A steady increase in average temperatures from global warming would dramatically alter today's pattern of tree distribution, putting northern coniferous forests under stress and encouraging the northward spread of deciduous species. In Finland, researchers are measuring the success of exotic tree species in Finnish conditions, and investigating the adaptability of present-day tree species to changes in climate. They have established a five-hectare (12-acre) arboretum and a gene pool forest with over 20 species of conifers and 20 species of hardwoods to maintain genetic diversity. Their aims are to find out how genes regulate tree characteristics, and to produce different varieties of seeds suitable for forest regeneration throughout the whole country.

Forests of the future

One of the most highly developed industrial forest plantation systems in the world is in Brazil. Between 1965 and 1985, more than 5.5 million hectares (13.5 million acres) of woody crop plantations were established, mainly in southern Brazil. (That's an area about the size of West Virginia.) Fast-growing pine and eucalyptus were introduced to these areas to compensate for the disappearance of indigenous trees. The government-sponsored program created a major new wood supply based on these short-rotation species, which increased rural employment and made Brazil a net exporter of wood products. The Brazilian companies involved in the program carry out research to improve the yields, quality, and sustainability of their plantations, and their results are spurring interest from tropical countries around the world.

The big criticism of industrial plantations is their massive impact on the environment through excessive use of chemical fertilizers, pesticides, and herbicides. And if, in the long run, they deplete the soil, these plantations may be unsustainable. Companies are meeting these criticisms in a number of ways, such as:

- adding leguminous species to plantations to improve soil fertility and keep down weeds, reducing the need for both fertilizers and herbicides
- leaving chipped logging residues and bark in the field to reduce nutrient loss and act as a mulch, which also decreases weed growth and the need for burning to prepare sites for the next planting
- using biological control against insect pests instead of using pesticides.

Environmental concerns combined with economic pressures from international trade agreements have forced forest managers to produce more crops more cheaply with less environmental harm, and in some cases biotechnology provides the tools to make this possible.

The ability to use bamboo, for example, as a raw material to make ethanol (see Chapter 5) has led to plans to expand bamboo planting in Brazil from about 15,000 to 60,000 hectares (37,000 to 148,000 acres) in the near future. The bamboo need not even take up more land, as it can be intercropped with food crops, and bamboo residues will provide pulp and livestock feed.

Closing thoughts

Bountiful and boundless — the earth's oceans and forests were described in terms like these for most of human history. Today, such words have a hollow ring. The mighty oceans grew less mighty when aircraft came continent-hopping into our world, and the forests that so recently clothed the tropics and northern continents are torn by gaping holes. On the brink of a new millennium, our rosy view of the earth's unlimited abundance has come to a gloomy end. Once-rich fisheries are abandoned, and forests are turned to barren landscapes or short-lived pastures. Our past progress now seems only a sorry history of plunder: generation after generation not caring to raise fish or plant trees since there was "always plenty more where they came from."

The old images of an infinite ocean and unending forest can never be real to us again, but there may be renewed hope of wealth in seas and trees. With the insights of biotechnology, they could have a future not only in fish and wood, but in new materials, medicines, chemicals, and fuels.

Knowledge is power, wrote Francis Bacon 400 years ago. Strange to discover that the long course of western scientific thought has come to this late point in its history before realizing how very little we know about two of the biggest ecosystems on the planet. How many unknown treasures still lie hidden in neglected corners of the ocean floor or in forest canopies — places of only academic interest to few people until recently? The true harvest of the oceans and forests in the next century will be knowledge, if only we can stop the destruction before it's too late.

Chapter 7
Ethical Issues

Should we, or shouldn't we? Ethical questions deal with the effects our actions (or inactions) have on the world around us. If something is harmful, we shouldn't do it. A simple enough guideline in theory, but one that's not very useful when the consequences of an activity aren't yet clear, or when its effects can be both harmful and helpful. Biotechnology falls into this ambiguous camp. Most differences of opinion between supporters and opponents of genetic manipulation come down to different interpretations of the balance between risks and benefits.

Typical concerns can be divided into a number of areas, ranging from biotechnology's effects on the environment and human health to impacts on social and economic conditions and religious and moral values (see the following page). Some issues arise specifically from the nature of the technology, while others, such as the exploitation of poor nations' resources by rich ones, are part of an existing dilemma.

Examples of biotechnology issues of public concern

Environmental safety

- Will genetically altered organisms upset the balance of populations in natural ecosystems?
- Will modified organisms transfer their altered genes to wild relatives or reduce biodiversity?

Food safety and health

- Will food from modified crops or livestock be safe to eat?
- Will genetically altered food have less nutritional value?

Social and economic effects

- What effects will biotechnology have on the business of farming around the world?
- Will patent laws give control of key crops to a few large companies?

Ethical and moral issues

- Are poor countries being exploited for their genetic resources?
- Do we have the right *not* to use biotechnology if it helps treat diseases or increase food production?

Regulatory issues

- Do current regulations give enough protection to farmers, consumers, livestock, and the environment?
- Should producers be required to label genetically altered food products?

Public perceptions of the facts of biotechnology, and the nature of the risks, are crucial to developing a consensus among science, public policy, and commercial interests. How do people first learn about biotechnology, and how do they react to it?

Making opinions

The average person gets news of biotechnology mainly through the media. Who provides this information? Does it present a balanced perspective? Does it deal with the various concerns people have about biotechnology?

To answer these questions, researchers at the Center for Biotechnology Policy and Ethics at Texas A & M University analyzed 132 newspaper articles about biotechnology, collected from a variety of newspapers throughout the United States during 1991 and 1992. The bulk of information (about three-quarters) quoted in the newspaper articles came from industry and university sources. Government spokespeople and groups opposed to biotechnology each supplied less than 10 percent of the information in the articles. The dominant users of biotechnology — farmers and physicians outside of research institutes — were very rarely cited. The news, in other words, was mainly reports of new discoveries, presented by the people who discovered them.

Arguments about biotechnology presented in the clippings focused on economic and health benefits, regulatory issues, and dangers. Industry spokespeople naturally emphasized the economic agenda, but were also more likely than university sources to talk about regulations and risks. In fact, comments about the potential dangers of biotechnology were as likely to come from industry sources as from critics of biotechnology. None of the newspaper articles were wholly negative, and arguments about public awareness or ethics were rarely reported.

The biggest boosters of biotechnology turned out to be universities. Academic researchers overwhelmingly tended to argue for the benefits of biotechnology, their positive comments outnumbering negative ones by three to one. Overall, universities painted a more one-sided picture than the biotech industries, which are assumed to have a strong vested interest in creating favorable public opinion.

The "bias" shown by universities may in fact have been a bias on the part of journalists, the researchers point out. Since industry and activists are seen to have a clear agenda at the outset, journalists may be more likely to ask these sources to justify what they say, in order to get balance for their articles. On the other hand, journalists might perceive university scientists as more objective and accept their comments at face value. As well, journalists with little scientific training may be intimidated by technical information or ill equipped to know what questions to ask.

What readers get in their newspapers is only one part of the equation. Do they get the kind of information they want? The study went on to compare the newspaper coverage of biotechnology with the concerns expressed in a previous survey of readers, and found some major discrepancies.

While nearly half the newspaper arguments focused on economic issues, only five percent of the survey group were concerned about this. Granted that at least some of the articles were intended for a business or investment audience, this emphasis by the newspapers seems out of keeping with the needs of their readers.

On the other side of the coin, more than one-quarter of the readers surveyed said they wanted more analysis of the news in the articles, to help them assess how much risk or benefit a particular piece of biotechnology posed. Nearly 10 percent of the survey group wanted to see more discussion of ethical issues, which appeared in only 1.3 percent of the newspaper articles.

Reviewing their results, the researchers concluded that "a media diet of boosting biotechnology and down-playing difficult questions may serve in the long run to promote cynicism and undermine public confidence...." The media, in short, don't appear to present the diversity of views and opinion needed to stimulate debate and develop consensus.

Direct surveys of public opinion about biotechnology have been carried out by a number of researchers in different countries, with broadly similar results. On the subject of gene transfer between species, popular judgment ranks living things in a hierarchy: half those surveyed find plant-to-plant transfers acceptable, but only about one in ten people feel comfortable about gene transfers between humans and other species. In general, people consider genetic manipulation of plants less troublesome morally than genetic manipulation of animals.

On widely publicized issues, such as the controversy over BST, people are skeptical. First and foremost, they want more information, and distrust what they hear both from industry and anti-biotechnology groups. Regarding milk and other food products made using altered genes, the great majority of people want to be informed by labels, such as those identifying organically grown food, so they can make a choice.

A 1993 report on public attitudes carried out for the Canadian Institute of Biotechnology looked not only at what people know and believe, but at the reasons why people tend to support or oppose biotechnology. People's attitudes on the subject are affected by their views on religion, science, and nature, by their perceptions of personal benefit, and by their trust in the process of decision-making by government.

The report identified three broad segments of society with different attitudes. One group, about one-quarter of those in the survey, felt that biotechnology offers more benefits than dangers to society. These people generally have faith in the ability of science and technology to solve problems, and are least likely to believe that nature is fragile, or a reflection of a divine will.

A second group, also one-quarter of those surveyed, felt that biotechnology offers more dangers than benefits. They are significantly less likely to believe that science is a way to "truth" and they mistrust the technological establishment. People in this group are most likely to see the world as a manifestation of "God's plan" and feel that modern technology is responsible for environmental crises.

The third group, representing about 40 percent of the public, believe that biotechnology is equally beneficial and dangerous to society. They are less extreme in commitment to either science or religion and prefer more citizen involvement in decision-making on these issues. The majority of these middle-of-the-road citizens are liable to weigh each issue as it comes, and to change their opinions on biotechnology according to specific cases.

Life®

A Questionnaire

Check the box that most closely reflects your response to each statement.

1. Corporations should be allowed to own exclusive rights to any genetically engineered organisms they develop.
   - ❑ strongly agree
   - ❑ agree
   - ❑ don't know
   - ❑ disagree
   - ❑ strongly disagree
2. Human genes should never be commercialized.
   - ❑ strongly agree
   - ❑ agree
   - ❑ don't know
   - ❑ disagree
   - ❑ strongly disagree

Whatever your opinion about the two statements above, it's a bit academic already. That's because corporations *do* own patents on organisms they've "invented" and human genes *have* been sold and bought.

Today's version of the successful inventor isn't the one who builds a better mousetrap, it's the one who builds a better mouse. The first patent for genetically modified mice was approved by the United States Patent Office in 1988. Created by geneticists at Harvard Medical School, the mice carry cancer-causing genes inserted into their cells and are used in studies of cancer development, and for screening anticancer drugs. The patented mice were commercialized by Du Pont in 1989 and sold under the trade name OncoMice.

The decision to grant a patent protecting a line of genetically altered animals was controversial. Originally, the patent system was developed to protect mechanical inventions. It later expanded to accommodate electrical and chemical devices and products. But organisms are both more complex and less predictable than physical systems, and have the disturbing ability to reproduce themselves with little or no help. Quite aside from the ethics of "owning" a line of animals or plants, how are you going to stop people from making their own copies once they've got a sample? Should offspring also belong to the original inventor, or just to the breeders who raise them?

The history of patent laws for organisms goes back to 1930 in the United States, when the Plant Patent Act gave growers rights to any novel strains of asexually produced plants they developed. The field was expanded in 1970 by the Plant Variety Protection Act to cover new varieties of sexually reproducing plants, but excluded their seeds. At the time, the possibility of altering single genes or switching genes between species was still not envisaged as a likely basis for commercial production.

A new era in patent history began in 1980 when the U.S. Supreme Court ruled that a patent could be granted for a bacterium genetically tailored to digest oil slicks. It's a measure of how rapidly biotechnology developed to see that the Harvard mouse came on the scene only eight years after that first landmark patent for an engineered organism.

As more applications to patent microbes, plants, and animals dropped onto U.S. Patent Office desks throughout the 1980s, the organization didn't bat an eyelid. On the contrary, it seemed remarkably easygoing, approving some broad claims that gave individual companies monopolies over entire areas of biotechnology and entire groups of organisms.

In Europe, views were initially different. A 1961 agreement among 17 nations allowed breeders to own varieties they created, regardless of who owned the parent stock. And in 1973, the European Patent Convention barred patents on animals and plants and on processes for producing them. But commercial pressures to impose common international patenting standards slowly pushed Europe towards the American view of things.

The European Parliament continued in its wish to exempt farmers from limits on their use of patented crops and livestock and to exclude human genes and tissues from patenting. However, the Parliament was opposed by the European Council, which sought to overturn that policy in favor of one with practically no limits on what could be patented.

The controversy between the two bodies was taken to a Conciliation Committee, which tabled a directive in 1988 asking Parliament to:

- legalize the patenting of human genetic material and gene therapies
- prevent farmers from freely reproducing patented livestock and seeds
- allow companies to own species of plants and animals
- allow patenting of genetic sequences and other basic discoveries.

The proposal was not adopted after a vote in March 1995, when there was still no agreement between the European Parliament and the Council. In the meantime, the European Patent Office (EPO) went on granting patents for inventions involving genetic engineering, including rights to human genes. All the while, an appeal board kept hearing arguments from opponents.

Patenting people

In an unprecedented move, on March 14, 1995, the U.S. Patent Office issued Patent No. 5, 397, 696 to the National Institutes of Health (NIH) for the genetic material of a foreign citizen, a Hagahai man from the highlands of Papua New Guinea. The Hagahai, who number only about 260 people, first came into regular contact with the outside world in 1984. The NIH now claims ownership of a cell line containing the man's unmodified DNA, together with several methods for using it to detect HTLV-1-related retroviruses.

"In the days of colonialism, researchers went after indigenous people's resources .... But now, in biocolonial times, they are going after the people themselves," says Pat Mooney, executive director of Rural Advancement Foundation International (RAFI), a group that leads opposition to the commercialization of human genes.

The NIH has sought patents on human genes in 19 other countries, usually with no concrete provisions to pay the original owners of the cells they take and use. Their interest is part of the Human Genome Diversity Project (HGDP), an international program that aims to sample blood and tissues from as many indigenous groups in the world as possible. Angered by what they dub the "vampire project," indigenous people, governments, and non-governmental organizations from across the South Pacific are working to establish a Lifeforms Patent-Free Pacific Treaty.

The main value of human DNA from remote populations is its potential to help researchers diagnose and treat diseases and develop vaccines. For example, blood samples drawn from asthmatic inhabitants of the remote South Atlantic island of Tristan da Cunha were sold by researchers to a California-based biotech company. The Californian company in turn sold rights to its still-unproved asthma treatment to the German company Boehringer Ingelheim for $70 million.

American claims for rights to human genetic material are pursued abroad by a division of the U.S. Department of Commerce. Clarifying his government's stand on this controversial issue, the former U.S. Secretary of Commerce, Ronald Brown, explained: "Under our laws . . . subject matter relating to human cells is patentable and there is no provision for considerations relating to the source of the cells that may be the subject of a patent application."

RAFI believes that this is the beginning of a dangerous trend in which indigenous people around the world are viewed as raw material for companies in the United States and other industrial nations. The organization has been monitoring the patenting of DNA from indigenous people since 1993, and is pressing international bodies and governments to bring the issue before the World Court at The Hague.

Scientists are obtaining genetic samples from isolated

populations to preserve a record of human diversity and evolution before these rare groups disappear into history. But opponents fear that the discovery of useful genes will inevitably lead to the patenting and marketing of portions of the human genome, an outcome they attack as exploitive and immoral. Writing to the National Science Foundation about the Human Genome Diversity Project, Leon Shenandoah of the Council of Chiefs of the Onandaga Nation proclaimed: "Your process is unethical, invasive, and may even be criminal. It violates the group rights and human rights of our peoples and indigenous peoples around the world. Your project involves the very genetic structures of our beings."

In Europe, the issue came before an appeal board when the Green Party of the European Parliament opposed an EPO decision to grant a patent for a human DNA fragment encoding a particular protein. Opponents argued that the DNA code was a discovery rather than an invention, and that giving a patent for a human gene offends morality.

The appeal board ruling dismissed both claims. On the first point, EPO guidelines permit natural substances to be recognized as novel when they are isolated for the first time. On the question of morality, the board ruled that the mere act of taking human tissue was not, as claimed, "an offense against human dignity" if the person from whom the tissue is taken consents. Taking tissue samples is standard practice in medical procedures. Nor can the patenting of human genes be considered "a form of modern slavery," since a patent to genes does not give any rights over the person from whom the genes were taken.

On the argument that the patenting of human genes is inherently immoral and tantamount to patenting life, the board's position was that the only thing being claimed was a particular chemical substance. The board agreed that "the patenting of a single human gene has nothing to do with the patenting of human life. Even if every gene in the human genome were cloned (and possibly patented) it would be impossible to reconstitute a human being from the sum of its genes."

The board found no moral distinction between "the patenting of genes on the one hand and of other human substances on the other, especially in view of the fact that only through gene cloning have many important human proteins become available in sufficient amounts to be medically applied."

Problems with patents

In order to get a patent, an invention must be novel, useful, and not obvious. The purpose of a patent is to give whoever holds it a number of years (usually 15 to 20) to have exclusive control over what they claim to have invented. Patent holders can then either monopolize production of their invention or license it to others.

The dilemma faced by biotechnologists is to know exactly what to claim from the results of their work, and at what stage in their research to file for a patent. In rare cases, the value of a discovery is clearcut and obvious — for example, a method for splicing DNA. More often, however, the applications of a new discovery are less clear, less of a breakthrough than an increment in knowledge.

To safeguard the potential value of their work and avoid losing the race to a competitor, some labs have been tempted to file broad patent applications at an early stage of their research. The U.S. NIH, for example, applied for patents on several thousand partial human DNA sequences it had identified, without knowing their functions or what commercial applications might result. Their application was rejected and they later abandoned it.

Others claim rights by extrapolation. Harvard's success with altering a mouse genome, for example, led them to claim their invention would apply to other mammals, even though they had not actually demonstrated this.

Although such claims may appear simply greedy and unreasonable, companies argue that they need the protection of a patent to repay the cost of their research and development. It typically takes several years and millions of dollars to bring a biotechnology application to market. If a patent application is limited to the very narrow and specific details of what has been achieved in the lab, it may not be enough to produce the profits needed to pay for all the work and capital invested. But on the other hand, if a patent claims a very broad area, such as a concept, a technique, or a group of plants or animals, it may restrict other researchers in the same field, slow progress, and divide the industry.

For example, an especially controversial decision in 1992 gave the American biotech company Agracetus a patent for *all* genetically engineered cotton plants. Scientists working for the company were the first to modify the genome of cotton using a bacterial species. On the basis of this process, they claimed patent rights to any transgenic cotton plants, no matter what the actual techniques used or genes altered. As one report put it, it was as if Henry Ford had been given a patent for all automobiles.

Agracetus obtained patents in other cotton-producing countries such as India, Brazil, and China, before the issue became a matter of wide public concern. The Indian government soon came to realize that the patent would deny Indian scientists the opportunity to develop their own varieties of pest-resistant cotton using recombinant DNA techniques. The Indian government revoked their license to Agracetus in October 1994, but India still cannot export any new cotton plants to countries where Agracetus holds a patent.

One of the criticisms of issuing broad patents is that it creates possessiveness about basic information, reducing the relatively free exchange of ideas and data traditional among scientists. Some of the discoveries now registered in patent files are things that, in the past, would simply have been published in science journals. Although private companies have always been concerned about shielding their research results, the trend to secrecy by publicly funded scientists in government and universities is not necessarily in the public interest.

Another criticism of patents is that a great deal of the basic knowledge underlying biotechnology was developed using public funding. In addition, many of the innovations claimed are relatively minor. Commenting on the patent given for the genetically engineered oil-eating bacteria he developed, Dr. Ananda Chakrabarty told *People* magazine, "I simply shuffled genes, changing bacteria that already existed. It's like teaching your pet cat a few new tricks." Allowing for Dr. Chakrabarty's modesty, and his oversimplification of the case, it's debatable whether companies should be given the right to reap great rewards for small modifications of naturally occurring organisms.

In the new areas opened up by biotechnology, adjudicating the legitimate extent of patent claims is as much a matter of interpretation as precedent. As each company's lawyers try to get the broadest protection for their employers, patent issues may pass from the patent office to the law courts. Several broad patents have been withdrawn after appeals, and limits on patenting DNA segments, proteins, and entire organisms are still being developed. The box below lists biotechnology products that have already been patented in one country or another. The United States tends to allow a broader range of patent claims than other countries.

What can be patented?

- genetically altered microbes such as bacteria, fungi, algae, other single-celled organisms, and viruses
- newly discovered microbes, if the invention includes an aspect not found in nature, or excludes their use as found in nature
- techniques for genetically manipulating or using microbes, plants, or animals
- cell lines (genetically distinct cells and all their descendants produced by normal cell division)
- genes, plasmids, vectors, and other DNA fragments, defined by a technical feature such as a nucleic acid sequence or restriction map
- monoclonal antibodies
- proteins prepared by a genetic engineering process, if they have altered properties not present in previously known proteins
- plant, animal, and human genes

Profiting from the poor

The quest for unknown organisms with useful properties has sent many "bioprospectors" to the world's tropical forests, and prompted others to study the agricultural and medical practices of indigenous cultures. The fruits of their research can bring large profits to the few biotech companies that develop them into products, but the countries where the discoveries are made are unlikely to get much in return.

The transfer of valuable resources from poor countries to rich ones is nothing new. But biotechnology is adding further insult to injury. The global distribution of modified crop seeds and livestock, for example, reduces the diversity of food grown around the world, increases costs to farmers, and makes everyone dependent on a few large corporations for this most basic of commodities.

The patenting of plants and animals means that farmers must pay royalties to the patent holder each time they breed their stock. The traditional farming practice of saving part of one year's crop to use as seed for planting the following year at no cost is no longer even possible with many hybrid crops. These crops cannot be regrown, and the farmer is forced to buy a fresh supply of patented seed each year, together with the agrochemicals on which the seeds depend.

Fed up with the appropriation of resources and the imposition of agricultural systems that work against them, half a million farmers in India demonstrated at the offices of the giant agribusiness Cargill in October 1993. They were objecting to the patenting of seeds they had used for thousands of years, and protesting against the effects of the General Agreement on Tariffs and Trade (GATT). GATT's goal of minimizing obstacles to international trade is widely criticized as serving the financial interests of multinational corporations more than the economic and social interests of citizens in member countries.

Under the agricultural and intellectual property provisions of GATT, patented genetic material belongs to the patent holder (usually a corporation), no matter where in the world it originated. This means that indigenous farmers can lose rights to their own original stocks, and not be allowed under GATT to market or use them. Peasant farmers go unrewarded for the cumulative knowledge built up over centuries about what to grow and how best to grow it, while corporations stand to harvest royalties from Third World countries estimated at billions of dollars annually. The very profitability of patented seeds makes it likely that companies will promote them in preference to older stocks, reducing the diversity of crops even further.

Much the same situation applies to the pharmaceutical industry. The healing potential of plants used by indigenous people may end up providing profits to drug manufacturers as a direct result of their patent rights, while people in poor nations where the plants are found cannot afford basic medical care. The industry argues that a patent is necessary for them to invest in the development of new drugs, whose production benefits everyone. The price of patented drugs, however, is often artificially inflated due to the monopoly, putting them out of reach of many people and increasing health insurance costs.

Protecting consumers

People want to know what's in their food and how it is produced. Some food labels address ethical concerns. For example, consumers want labeling of environmentally friendly and socially responsible products, such as tuna that have been caught without killing dolphins in fishing nets. Other labels are important for health reasons. But whatever the issue, labeling with relevant information at least allows consumers to choose whether to buy a product or not.

People with food allergies are particularly concerned over transgenic foods, since a chemical to which they react badly may be transferred by genetic engineering to a food in which it was previously absent. For example, some people have an inherited metabolic deficiency named favism, which causes them to react adversely to the seed protein lectin, found in legumes such as beans. These people avoid eating beans. Lectin, however, deters aphids from feeding on legumes, and the gene for making lectin has recently been engineered into potatoes as a pest defense strategy. The risk is that individuals with favism may unknowingly eat these transgenic potatoes and suffer as a result. Accurate labeling is their only defense against such a possibility.

Another health concern is that transgenic food carrying marker genes for antibiotic resistance might transfer the resistance to consumers eating the food. This raises the risk that resistant genes might be incorporated into germs, against which we would then have no defense. To reduce the use of antibiotic genes, researchers are developing other genetic markers, based on such things as color change, or ability to use certain sugars.

At present, there are no regulations requiring the separation of transgenic livestock and crops from other animals and plants. As a result, some vegetarians have expressed concern they might eat vegetables to which animal genes have been transferred. Some religious groups have questions about breaching dietary prohibitions, and some have even suggested that eating, say, pigs that have been modified by human genes makes consumers into cannibals.

These last arguments are not very persuasive. As I explained in Chapter 2, DNA is not species specific — which is the very reason genes can be swapped among different organisms. We already have genes in common with many other species of animals, plants, and microbes. To say that a plant with an animal gene in it has an animal component expresses a fundamentalist view of species that doesn't exist in nature. One might as well argue that a plant whose roots take up minerals from a decomposing animal has an animal component in it.

Health dilemmas

Would you like to be told that you are very likely to develop an incurable disease within a few years? Should a doctor automatically give such information to a patient? Should a patient share such information with family and friends? Genetic screening has raised this issue by giving doctors the ability to diagnose genetically related (but still untreatable) disorders. On one hand, such knowledge can create depression and anxiety in patients; on the other hand, it can help them prepare for the disease and receive counseling.

With the increase in numbers of metabolic and genetic disorders that can now be diagnosed, the practice of genetic testing and screening has greatly increased in recent years. Screening in the United States for cystic fibrosis (CF), for example, jumped from just over 9,000 tests in 1991 to 63,000 tests in 1992. The U.S. President's Commission for the Study of Ethical Problems in Medicine and Biomedical and Behavioral Research predicted in 1983 that genetic screening and counseling would become major components of health care in that country by early in the 21st century.

In some cases, prenatal and newborn screening can help detect genetic diseases for which there is some remedy. For example, phenylketonuria (PKU) is a rare metabolic disorder that causes mental retardation, but its effects can be prevented by following a special diet. In most cases, however, screening doesn't reduce the incidence of illness or death because the illness does not yet have a treatment.

Apart from posing individual dilemmas to both doctor and patient, genetic screening opens questions of privacy, stigmatization, and effects on employment and insurance. Will employers, insurance companies, or police have access to an individual's genetic information? How are genetic differences to be described? Terms such as abnormality, flaw, or defect could lead to discrimination. In court cases, how reliable are the genetic tests used to link suspects to a crime?

A new use of screening is to identify workers who might be particularly susceptible to substances found in their workplace. Thousands of workers are disabled by occupational illness each year. While it seems like a useful tool to protect vulnerable workers, this use of genetic screening involves some degree of crystal ball gazing, and might reduce job opportunities for people who, at the time of the test, are healthy and able.

The Committee of Ministers of the Council of Europe looked into these issues and recommended that genetic screening be used with caution. It was important, they agreed, to have public education in advance. The Committee declared that screening should not, in any case, be compulsory. Insurers should not have the right to require genetic testing or to seek the results of previous tests. The Danish Council of Ethics pointed out in 1993 that genetic information reveals knowledge not only about an individual, but also about the individual's relatives. This creates more dilemmas over confidentiality concerning the potential risk of disease.

According to the Privacy Commission of Canada, genetic privacy has two dimensions: protection from the intrusions of others and protection from one's own secrets. It concludes that privacy of genetic information is an explicit constitutional right protected by legislation and should be used only to inform a person's own decisions. Employers should not be allowed to collect genetic information, and services and benefits should not be denied on the basis of genetic testing.

In human terms, the easy access of genetic screening might place people under pressure to be tested for all sorts of situations, from planning marriage, to traveling, starting a new job, or deciding when to retire. If screening comes to be seen as a social good for improving community health, there might be prejudice against those who decline to be screened. Individuals found to have certain genetic predispositions might come to see themselves as victims of fate, or be branded as "abnormal."

A 1992 poll of American citizens found that 68 percent of people questioned knew little or nothing about genetic testing, but 79 percent would undergo testing before having children to learn whether their child might inherit a fatal genetic disease. About three-quarters of people questioned favor strict regulations on the use of genetic screening.

Gene therapy revisited

U.S. PATENT NO. 5,399,346

CLAIM 1: "A process for providing a human with a therapeutic protein comprising: introducing human cells into a human, said human cells having been treated in vitro to insert therein a DNA segment encoding a therapeutic protein, said cells expressing in vivo in said human a therapeutically effective amount of said therapeutic protein."

That piece of legalese above is part of the patent for gene therapy issued in March 1995 to William French Anderson, Michael Blaese, and Steven Rosenberg — scientists whose research I described beginning on page 65. Because much of their work on this technique was undertaken for the U.S. National Institutes of Health, the patent rights were assigned to the U.S. government. The government then gave a private company, Genetic Therapy Inc. (GTI) of Maryland, exclusive license to commercialize the technology. Essentially, a medical procedure developed at public expense was privatized. This move generated controversy in industry, government, and university labs around the United States over the limits this might place on other labs and hospitals to carry out treatments on their patients.

The application for this patent took five years to review before it was approved, but as of writing the patent stands to be challenged by a number of sources. An important factor that remains to be seen is how readily GTI will sub-license others to use the therapy, and how reasonable its terms will be.

A common objection to patents of this kind questions the novelty of the invention. How much does it owe to unacknowledged previous work in the field? With research that involves teams of scientists and work carried out over several years, it's difficult to establish where credit for a new advance is due. For example, Martin Cline of the University of California at Los Angeles first tested gene therapy to treat thalassemia (a red blood cell disorder) in 1980, laying the groundwork for Anderson and his colleagues 10 years before their patent claim was submitted. Did Cline's work establish gene therapy as a viable technology? He hadn't received approval from UCLA for his clinical trials and conducted them outside the U.S.A. — in Israel and Italy.

But even if previous work had not already beaten the patent applicant to the punch, sometimes such research makes the subsequent invention obvious to those in the field. In this example, it can be argued that many people were still skeptical about prospects for the success of gene therapy at the time the patent application was filed.

A third objection is that certain developments crucial to the invention were created by other parties, who should be listed as co-inventors on the patent claim. For example, the retrovirus vectors used in one of the gene therapy methods played a significant role in its success.

All in all, the issues spilling from the patenting of gene therapy create a field day for ethicists. Problems range from the ethics of manipulating human genes in the first place to the rights and wrongs of restricting techniques designed to save human lives.

Long before the dust settles on any litigation that may arise, gene therapy will have moved onward and made some of the issues — and the patent — obsolete. Already, much of the research work in gene therapy now focuses on in vivo techniques. With these techniques, corrected genes are inserted directly into cells in the body, rather than into cells cultured in vitro, the process claimed by the patent. For example, researchers in the United States are now attempting to insert corrected cystic fibrosis genes into patients' lungs using aerosol sprays. In the fast-moving world of biotechnology, the 15 to 20 years of protection given by a patent may in the end leave an inventor high and dry as the stream of invention flows elsewhere.

Pros and cons of gene therapy

Should we be altering something as fundamental as people's genes? The issue still causes dispute 16 years after the first attempts were made on human patients. There are important distinctions to be made between altering the somatic genes found in most body cells (which affects only the person concerned) and altering germ-line genes found in sperm and egg cells (which affects descendants of the patient). A study published by the U.S. Congress Office of Technology Assessment in 1984 reported a consensus among civic, religious, scientific, and medical groups that, in principle, somatic-cell gene therapy is appropriate for humans.

Here is a summary of the arguments commonly used in favor of gene therapy:

- It may be the only way to treat certain disorders in desperately ill patients, or to prevent the onset of illness in others.
- Compared with the hardship and risk of death faced by these patients, the uncertainties of gene therapy are acceptable.
- We have an obligation to treat severe illnesses if we can.
- Prohibiting gene therapy research restricts the intellectual freedom of researchers.

In addition, the following points have been made in support of the more controversial issue of germ-line gene therapy:

- It offers a true cure for genetic illnesses, not simply a treatment of symptoms.
- By preventing the transmission of disease-causing genes, the risks and costs of therapy for future generations are reduced.
- Doctors are obliged to respond to the health needs of prospective parents who are at risk for transmitting serious genetic diseases.

Supporters of gene therapy research concede that it involves experimenting with human embryos, but argue that this is necessary to benefit generations in the future. Successful gene therapy will make it possible to save the lives of many infants who would otherwise die, and reduce the need to make difficult decisions about what to do with embryos that have a genetic disease.

But concerns about gene therapy persist, and here are some of the reasons why:

- It's the start of a slippery slope. Once the techniques of gene modification have been developed, they are open to misuse, tempting those in power to alter genes for reasons other than eliminating disease.
- The long-term effects of germ-line gene therapy can't be assessed without clinical follow-up of patients over generations, a difficult if not impractical prospect.
- The long-term implications can't be understood by the children who make up a large proportion of gene therapy candidates.
- Having a choice of whether or not to use gene therapy creates a conflict of interest between the reproductive liberties and privacy of parents and the interests of insurance companies and society over the financial burden of caring for children with serious genetic defects.

There are many other responses to the issues of gene therapy from individuals and organizations around the world, reflecting the complexity of this debate. Some of the arguments can be applied to the increasing cost of high-tech medical care in general, and the price society pays for a strained health care system. Can we afford such expensive therapy, and who should receive it? And who decides?

Prometheus revisited

Prometheus was the Greek demi-god who stole a spark of fire and was punished by Zeus for his presumption. In the minds of many people, the enterprise of biotechnology is a Promethean risk, another example of humanity's self-destructive aspirations to play god. But it's rather late in the game to object to human nature. We have benefited and suffered from our curiosity since the days when we discovered that rocks have better uses than to be left lying on the ground.

Powerful though our species has become, it is a mark of hubris to believe we *can* play god. For all our inventions, we do not, literally, create anything: we only take what nature provides and alter it for our own purposes. What, then, is meant by those who fear that biotechnology is "tampering with nature"?

Critics of recombinant DNA research during the 1970s focused on the risks that newly constructed organisms might pose — to human health, or the health of other species, or to important ecological processes. In short, they wondered if the products of biotechnology were safe.

Risk assessment is partly a matter of data, partly a matter of interpretation and temperament. For example, during the Cold War years, many people lived with the fear that the buildup of nuclear weapons would inevitably lead to nuclear attack, and prepared for such an event. Others believed, with equal plausibility, that the very power of these weapons was a defense that made the world safer. Was one view right and the other wrong?

Risk assessment is also affected by familiarity. We're generally more willing to live with familiar risks than new ones, no matter what the relative dangers. (Few people fear car travel, for example, despite the numbers killed daily on the roads. More people fear air travel, which is much safer.) Now that genetically altered bacteria have been handled for more than 20 years without disaster, earlier anxieties about mutant germs have diminished. Today's concerns are more likely to be about such things as the ethics of patenting genes and the exploitation of farm animals and indigenous people.

While we wait for tangible signs of harm caused by biotechnology, it must also be said that proof of danger doesn't necessarily lead us to abandon a particular activity. If it did, we'd outlaw cars and trucks tomorrow. Apart from the regular toll of death and injury from traffic accidents, gas-powered vehicles degrade our health and the environment, consume huge amounts of nonrenewable resources, distort land use planning and destroy neighborhoods, and arguably create a net drain on the economy. The fact that we'll be happily driving our vehicles tomorrow argues that the path of social change isn't built on principles of logic alone.

Because the benefits of cars are much more obvious to us than their hazards, we remain relatively uncritical of this technology despite the weight of evidence against it. With a new technology, such as genetic engineering, where we have as yet little or no personal experience of either its benefits or risks, the risks are more liable to occupy our imaginations.

Some of the fears I've read and heard expressed about biotechnology tend to come up time and again, and I'll try to answer them now from my own perspective on the subject.

Altering genes is unnatural. Genetic mutations occur naturally in all living things. They result from physical or chemical damage to DNA or from spontaneous errors that occur during cell division. Genes may also move from one place on a chromosome to another, and undergo duplication. Mutations and chromosomal rearrangement often have drastic effects, but they are part of the driving force of evolution, throwing up new characteristics on which natural selection can act. Without the alteration of genes, evolution could not occur.

Swapping genes between species is unnatural. Gene swaps between species are not entirely human inventions: they occur in nature. Among microbes, genetic exchange between species is common. Viral infections also carry genetic material from species to species, even among widely different organisms such as insects and mammals. Reproduction between closely related species of animals and plants is widespread, if uncommon in the wild. In captivity, such different types of animals as tigers and lions, or zebras and horses, have been persuaded to pair, and horticulturalists regularly crossbreed different varieties of plants. Hybrids may or may not be sterile, depending on the compatibility of the two species' genes.

Yes, but genetic engineering breaches fundamental species boundaries. Species are dynamic, ever-changing entities. The idea that rigid boundaries separate one from another is more a product of the human mind than of nature. Taxonomists — specialists in biology who describe and name species — frequently disagree over where one species ends and a similar one begins. At the molecular level, where genes function, boundaries are even less clear. Fundamental metabolic processes are similar in all living things, and biotechnological research overwhelmingly vindicates Darwin's thesis that species share much in common as a result of common origins. Even the apparently solid boundary between plant and animal kingdoms is perforated. For example, many species of plants have genes for producing animal hormones and enzymes, which they make as defenses against mammals and insects that feed on them.

New combinations of genes in microbes are likely to produce dangerous and uncontrollable mutant germs. The possibility of genetic engineers inadvertently making dangerous new microbes does exist, but it is a small risk. Disease-causing organisms are often very specialized, and microbes engineered in the lab for particular purposes are unlikely to outcompete their wild relatives if they should "escape." Virulent diseases such as AIDS, flu, bubonic plague, and Ebola all developed naturally, and genetic engineering is unlikely to produce anything worse. Another possibility is that genetic engineers could deliberately create harmful new germs, but those who choose to develop germs as weapons can do so with or without biotechnology.

We are altering evolution by creating new combinations of genes. We have created new crops, livestock, and domestic pets for centuries by altering wild ancestral genes through selective breeding. Far from breaking away and putting their mutated, domesticated genes back into wild populations, most of these altered organisms have become ever more dependent on people for their survival. We have had a much greater impact on the course of evolution over the centuries by transporting animals and plants from continent to continent. Introduced to places where they did not originally evolve, hardy organisms such as rabbits, starlings, rats, and various weeds spread rapidly and become pests, displacing native species or driving them into extinction.

Biotechnology brings unprecedented new power to humanity, with new ethical dilemmas. At the root of this view lie fundamental questions about humanity's place in the universe. Biotechnology is a tool that can be viewed from many perspectives. It has the potential to bring benefits and dangers. It requires safeguards and new laws and regulations. It is open to abuse. But that much can be said about other recent technologies, such as the Internet, cellular phones, and contraceptive pills. Does biotechnology add distinctly new challenges, or only different versions of the same old ones?

Most people do not see life merely as a collection of chemicals. In appearing to view life that way, the discoveries of biotechnology may seem cold and dehumanizing to some. But science is not to blame for society's lack of spiritual values, any more than literature or music is to be blamed for not feeding the world or curing diseases. Science is merely a way of understanding how the world works, and in this it has been described by Sir Peter Medawar as "incomparably the most successful enterprise human beings have ever engaged upon."

Revelations about the structure of nature can never be harmful to humanity, although knowledge of molecular biology, like any knowledge, can be misused. Nor does science pretend to say all that is important to people in their lives. Ironically, however, disillusionment with science and technology is on the rise at the very time when understanding of science is most needed to make important decisions for society's future. In their book *Reshaping Life: Key Issues in Genetic Engineering*, Australian authors and scientists G.J.V. Nossal and Ross L. Coppel write: "In the deepest sense, DNA's structure and function have become as much part of our cultural heritage as Shakespeare, the sweep of history, or any of the things we expect an educated person to know."

The fears I outlined briefly above are variations on the theme of "tampering with nature." But it's hard to see how biotechnology's tampering is any worse than the tampering we've already done with conventional, even mundane, technologies. Whenever we build dams, cut forests, drain wetlands, irrigate deserts, mine ores, build cities, expand agriculture and fisheries, and pollute the environment, we put our stamp on the face of nature around the planet.

The expression "tampering with nature" implies that people are somehow outside of the rest of the living world, a self-awarded status that has been a part of our cultural tradition for many centuries. My own feeling is that the revelations of biotechnology add significantly to the view that our own species, for all its uniqueness, is not fundamentally different from others. The emperor has no clothes that we can see, and I believe it is that which people object to most.

The issues of how we treat one another and the world around us in a responsible and ethical way cannot be divined through nature. They are human issues and depend on human values. The choices have always been with us, and were not created by biotechnology. It has always been possible for us to use our technology for good or bad, since the time when *Homo sapiens* first picked up a stone from the ground. That much has not changed.

Postscript

I have tried to put my finger on what it is about biotechnology that causes some people to hold it out as a great hope for the future and others to reject it as a dangerous and unwise application of science. How is it that adversaries can look at the same evidence and come up with opposite interpretations? Is the evidence too inadequate to be conclusive? Or is something else going on?

One answer to the dilemma is another apparent paradox: both sides are right, and neither side is right. Like many technologies, biotechnology has both its advantages and disadvantages. When opponents focus on different, if overlapping, issues, the result is confusion and disagreement. Used as a catchall term to cover a multitude of effects, biotechnology often ends up getting sole blame or credit for outcomes that are really due to a combination of factors that include planning, infrastructure, regulations, and economics. The way to resolve these disputes is to be more precise about what exactly is being debated in the name of biotechnology.

For example: Does the Human Genome Project provide knowledge that can help prevent and cure disease? Yes. Does it give rise to difficult ethical dilemmas (such as medical privacy and the patenting of human genes)? Yes. Should we abandon the project because of the difficulties? Or, how can we control and manage the project so as to minimize the ethical concerns and reduce known risks while keeping the benefits? The last question is the most difficult to answer. It is less likely to be addressed in public debates, which tend to be framed in simpler terms of yea or nay.

The main arenas in which biotechnology battles are fought — medicine, agriculture, and the environment —

add to the confusion. They come already littered with philosophical and ideological debris, which combatants pick up as ready-made shields or weapons. Holistic medicine, industrial agriculture, vegetarianism, free-market economy, corporate control, and consumer lifestyles all become part of the arsenal, and the resulting melee quickly obscures the particular details of biotechnology and its tools.

Once we get down to specifics, the path may be straighter but the going may be no less rough. For example, is it a good thing or a bad thing to add genes for growth hormones to salmon? Let's see:

1. Engineered salmon grow bigger faster on less food, which is a good thing economically.
2. They contain no new hormones (only more of the fish's own hormone), so there is no reason to anticipate health risks to consumers.
3. But the fast-growing fish might be more prone to disease.
4. And if they escape from fish farms into rivers and lakes, they might harm wild stocks by competing for food or spreading diseases.

Of the four points made (and more might be added), two are positive and two negative. But the degree of certainty of each statement declines as you read down the list. This example is fairly typical. We can often be more sure about the economic outcomes of biotechnology than we can about health or environmental outcomes. And that, to many people, is a big problem.

Uncertainty is a stock-in-trade of the prediction business. The inherent complexity of the interactions that produce healthy people and a healthy environment will always limit our ability to guarantee certain outcomes. There will always remain the conundrum that we cannot know the risk of releasing genetically engineered organisms into the environment without actually doing it. The best that can be done in advance is to carry out studies in confined situations that simulate the natural world, and that is all we can expect and demand.

Uncertainty is also, in part, a product of our own ignorance. It is important to remember how new many techniques are. A geneticist in the 1960s would have known nothing of the basic techniques used today to manipulate genes. In 20 or 30 years, our knowledge will be greater, and our uncertainty in some matters will be less. Of course that's no comfort to those who are already convinced that the outcomes of biotechnology are likely harmful.

Taking stock of the situation from a scientific point of view, I'd say there's little evidence to date that biotechnology itself has resulted in any significant harmful effects to the environment or human and animal health. There have been glitches and disappointments, but no mutant organisms out of control, no epidemics, no environmental catastrophes. There have, on the other hand, been several important benefits from biotechnology, including new methods of treating disease in people, crops, and livestock, and of cleaning up environmental pollution. From this perspective, I would say that biotechnology has been a net asset, if only on the general principle that it is better to have a wider choice of tools to use than a smaller choice.

Before anyone assumes that this conclusion puts me in a pro-biotechnology camp, I'll repeat that my view is based on the scientific evidence I've read, and on things that have actually happened, rather than things that have been predicted. It is not the only perspective. There's a social aspect to biotechnology, and there the evidence is more equivocal.

Biotechnology grew initially out of the desire of scientists to find out how the world works. Facts about genetic structure and the mechanisms of genetic control were first discovered by university researchers pursuing their scientific curiosity about nature. Their goals were to know such things as why one cell is different from another, or how DNA replicates itself. Probably few if any of those who laid the basic groundwork were motivated by thoughts of using microbes to make drugs.

The context in which most biotechnology research is conducted, however, is very different. Biotechnology is first and foremost a commercial activity — a reality that largely determines the priorities and goals of what is investigated and how it is applied. While the world at large may lack adequate vaccines, food, and pollution control, the focus of biotechnology companies is profit, not philanthropy. The large sums of money needed for research ensure that products with maximum profit potential get priority for development. As Canadian science broadcaster David Suzuki put it in an interview: "Biotechnology research serves the desires of the rich rather than the needs of humanity."

A good example of the confusing links between pure and applied research, and between economics and ethics, came into the news as this book was being prepared for printing. The successful creation of a healthy lamb cloned from the udder cells of an adult sheep (announced in February, 1997) caused astonishment even among many biologists and prompted a host of alarming comments about the prospect of cloning humans.

The tremendous breakthrough in technique was achieved by Dr. Ian Wilmut and his colleagues at the Roslin Institute in Edinburgh. Their goal was not the academic one of finding out about the control of genes but the practical one of duplicating sheep, as a step towards developing engineered animals for making drugs. It was an exercise in pure animal husbandry technology that would have been unlikely to receive a grant from bodies funding research in human health — even though its final application includes the production of pharmacologically useful drugs.

It is a comment on the increasingly narrow specialization of modern scientists that the Scottish feat came as a complete surprise to some of the leading researchers in the high-profile fields of molecular genetics and reproductive biology. Many of the latter had concluded from their research with small lab animals that cloning from the differentiated body cells of adult mammals was impossible. Meanwhile, the Edinburgh team went on developing their company-sponsored knowledge in a world largely unnoticed by other scientists. The team was seemingly unmindful of the ethical issues that could arise if their work were applied directly to humans.

Anxieties about cloning people featured prominently in news reports of the research. Often portrayed as a "taboo" and as a "dreaded" result of genetic engineering, the genetic duplication of a human individual was presented as self-evidently undesirable, even though it occurs naturally every day in the form of identical twins. The precise concerns were rarely articulated. Perhaps on closer examination most of them might turn out to be either unjustified or little more than a kind of human chauvinism ("it can't be done to humans because we're humans").

The domination of scientific research by corporate interests is at the core of most anxieties about biotechnology. The rush to patent and the reluctance to label only add to the perception that companies are greedy and untrustworthy, despite protests from business and other interests that patents are needed to protect large investments and that labels are not meaningful.

The huge scope of the social issues and their ramifications have been dealt with in more detail in some of the publications I've listed for further reading. I believe that many public anxieties about biotechnology are misplaced and are best addressed by more education. Others have merit and must be dealt with by more stringent public controls and monitoring of industry. In most cases, however, the only honest conclusion possible is to concede that time alone will tell the truth of the matter.

Perhaps the greatest potential tragedy of all in this story is that the squabbling factions now fighting to control the body of science and technology may end up tearing it apart. Our researchers' and technologists' abundant skills and energy may be frittered away, like so many once-plentiful resources in recent times, and we will find ourselves in the painfully ironic situation of being able to solve our world's desperate problems in theory, but not in practice.

I must conclude by making a special plea for the role of scientists and science education in public policy. Freedom to explore is fundamental to the progress of science. That freedom is threatened today by a number of sources, from the narrow agenda of companies, to government indifference and cutbacks, to public skepticism, fear, and lack of support. If, as I believe, the truth will out, it can only do so if the pattern of scientific research is not distorted by one vested interest or another.

Glossary

alkaptonuria: a genetic disease in which the urine turns black when exposed to air, due to homogentistic acid in the urine.

amino acids: naturally occurring biological molecules with a variety of functions. Among the amino acids, there are 20 that are used as building blocks for making proteins.

antisense therapy: administration of a drug consisting of short pieces of artificially produced, single-stranded DNA (about 15 to 25 nucleotides). The DNA is complementary to a section of an RNA molecule. It base-pairs with the RNA and prevents it from making a protein that is harmful to the system.

***Bacillus thuringiensis*:** a strain of bacteria that produces a protein toxic to certain insects that cause significant crop damage. The bacteria are often used for biological pest control. Recently, the gene that codes for the toxic protein has been engineered into other soil bacteria and also directly into some crop plants.

bacteria: one of the five kingdoms of living things. Bacteria are structurally simple single cells with no nucleus.

bacteriophage (or phage): a virus that infects bacteria. They are used by genetic engineers to introduce genes into bacterial cells.

base: one of the building blocks of DNA or RNA. A nitrogen-containing base combines with sugar and phosphate molecules to make a nucleotide. The four bases in DNA are adenine (A), guanine (G), cytosine (C), and thymine (T).

base pair: two nucleotides held together by a weak bond between complementary bases. In DNA molecules, adenine is paired with thymine and guanine is paired with cytosine.

cells: the basic structural units of life.

chromosomes: threadlike bodies that carry the genes. They can be seen in the nucleus of a cell just before it divides in two.

clone: a collection of genetically identical copies of a gene, cell, or organism.

codon: a triplet of nucleotides that is part of the genetic code and specifies the particular amino acid to be added to a growing chain to make a protein.

cyclosporine: a drug produced by a soil fungus; it is used to prevent organ rejection by inactivating the body's T-cells.

DNA: deoxyribonucleic acid. The genetic material of organisms (except retroviruses), made of two complementary chains of nucleotides wound in a helix.

gene: the physical unit of inheritance, made up of a particular sequence of nucleotides on a particular site on a particular chromosome.

gene expression: the conversion of the gene's nucleotide sequence into an actual process or structure in the cell. Some genes are expressed only at certain times during an organism's life and not at others.

genetic code: the sequence of nucleotides in a gene, coded in triplets (codons). The genetic code determines the sequence of amino acids in protein synthesis.

genome: all the genes in a complete set of chromosomes.

hirudin: a potent clotting inhibitor produced by leeches. The gene for this protein has now been genetically engineered into canola plants.

Human Genome Project: an international research effort begun in the 1980s to map and sequence all 100,000 or so genes found in human DNA.

hybridoma: a fast-growing culture of cloned cells made by fusing a cancer cell to some other cell such as an antibody-producing cell.

integrated pest management (**IPM**): the use of combined strategies to combat pests, including chemical, physical, and biological methods of control.

metallothioneins: protein molecules that bind specifically onto certain metals.

monoclonal antibody: antibody of a single type produced by a genetically identical group of cells (clone). Usually a fusion of an antibody-producing blood cell and a cancer cell. See hybridoma.

nucleotides: a compound consisting of a base, a phosphate group, and a sugar. DNA and RNA are linear chains (polymers) of nucleotides.

nucleotide sequence (**or base sequence**): the particular arrangement of nucleotides along a strip of DNA. Genes are defined as a particular nucleotide sequence.

nucleus: part of the cell containing the chromosomes.

oncogenes: tumor-causing genes associated with cancer.

osmostic pressure: the pressure that develops when the water solutions on the two sides of a semipermeable membrane have different concentrations of dissolved materials.

periphyton: a thin layer of algae, bacteria, fungi, and other microorganisms found on submerged surfaces in fresh water.

phage: short for bacteriophage.

plasmid: a small circle of bacterial DNA, separate from the single bacterial chromosome, and capable of replicating independently. Plasmids are also occasionally found in certain fungi and plants.

polymerase chain reaction (**PCR**): a method for making multiple copies of fragments of DNA. It uses a heat-stable DNA polymerase enzyme and cycles of heating and cooling to successively split apart the strands of double-stranded

DNA and use the single strands as templates for building new double-stranded DNA.

proteins: molecules made up of long chains of amino acids. They build tissues and carry out many critical functions in the body. Proteins literally make us what we are.

recombinant DNA: novel DNA made by joining DNA fragments from different sources.

restriction endonuclease (or enzyme): an enzyme that cuts a DNA molecule at a particular base sequence.

Restriction Fragment Length Polymorphism (RFLP): the variation, within a population, of the lengths of restriction fragments formed by treating DNA with a restriction enzyme. Responsible for the difference in DNA fingerprints of individuals.

retrovirus: virus having RNA as the genetic material. Inside the infected host cell, the viral RNA is used as a template for making viral DNA, which then becomes integrated into the host cell's chromosomal DNA. From there, the viral DNA can direct the formation of more, identical viruses.

reverse transcriptase: a retroviral enzyme that uses RNA as a template for making DNA.

ribonucleic acid (RNA): A nucleotide chain that differs from DNA in having the sugar ribose instead of deoxyribose, and having the base uracil instead of thymine. RNA helps translate the instructions encoded in DNA to build proteins.

transgenic organism: an organism into which the genes of other species have been engineered.

vector: in genetic engineering, an entity used to carry recombinant DNA into a cell. Plasmids and phages are commonly used as vectors.

Further Reading

If you'd like to read more about topics covered in this book, check out the following recommended books.

Chapter 1. **How Biotechnology Came About**

Levine, Joseph and David Suzuki. *The Secret of Life: Redesigning the Living World*. Toronto: Stoddart Publishing Co. Ltd., 1993.

Prentis, Steve. *Biotechnology: A New Industrial Revolution*. New York: George Braziller, 1984.

Watson, James D. et al. *Recombinant DNA: A Short Course*. New York: Scientific American Books, 1983.

Chapter 2. **Tools in the Genetic Engineering Workshop**

Cherfas, Jeremy. *Manmade Life: A Genetic Engineering Primer*. Oxford: Basil Blackwell, 1982.

Drlica, Karl. *Understanding DNA and Gene Cloning*. New York: John Wiley & Sons, 1984.

Lee, Thomas F. *The Human Genome Project: Cracking the Genetic Code of Life*. New York: Plenum Press, 1991.

Trefil, James. *The Year in Science: An Overview. Encyclopedia Britannica Yearbook of Science and Technology*, 1989, 271-275.

Wallace, Bruce. *The Search for the Gene*. Ithaca, N.Y.: Cornell University Press, 1992.

Williams, J.G and R.K. Patient. *Genetic Engineering*. Oxford: IRL Press Ltd., 1988.

Chapter 3. **Biotechnology and the Body**

Kimbrell, Andrew. *The Human Body Shop: The Engineering and Marketing of Life*. San Francisco: Harper, 1993.

Lipkin, Richard. "Tissue engineering: Replacing damaged organs with new tissue." *Science News*. July 8, 1995, Vol. 148, 24-26.

Lyon, Jeff and Peter Gorner. *Altered Fates: Gene Therapy and the Retooling of Human Life*. New York: W.W. Norton and Co. Inc., 1995.

Platt, Anne E. "Infecting ourselves: How environmental and social disruptions trigger disease." *Worldwatch Paper 129*, Washington: Worldwatch Institute, 1996.

Chapter 4. **Biotechnology on the Farm**

Brown, Lester R. and John E. Young. "Feeding the world in the nineties." *State of the World 1990*. New York: W.W. Norton and Co. Inc., 1990, 59-78.

Carson, Rachel. *Silent Spring*. Boston: Houghton Mifflin Co., 1962.

Kaiser, Jocelyn. "Pests overwhelm Bt cotton crop." *Science*. July 26, 1996, 423.

Kneen, Brewster. *From Land to Mouth: Understanding the Food System*. Toronto: NC Press Ltd., 1993.

Paoletti, Maurizio G. and David Pimentel. "Genetic engineering in agriculture and the environment: Assessing risks and benefits." *Bioscience*. Vol. 46, No. 9 (1996), 665-673.

Chapter 5. **Biotechnology and the Environment**

Fincham, J.R.S. and J.R. Ravetz. *Genetically Engineered Organisms: Benefits and Risks*. Toronto: University of Toronto Press, 1991.

Levy, Stuart B., and Robert V. Miller, eds. *Gene Transfer in the Environment*. New York: McGraw Hill, 1989.

Tudge, Colin. *The Engineer in the Garden*. London: Cape, 1993.

Chapter 6. **Biotechnology in Seas and Trees**

Attaway, D.H. and O.R. Zabrosky, eds. *Marine Biotechnology: Pharmaceutical and Bioactive Natural Products.* New York: Plenum Press, 1993, 419-457.

Marine Biotechnology. Special issue published by *Bioscience.* April 1996. Vol. 46, No. 4.

Chapter 7. **Ethical Issues**

Anderson, Norman G. "Evolutionary significance of virus infection." *Nature.* Vol. 227, Sept. 26, 1970, 1346-1347.

Keevles, Daniel J. and Leroy Hood. *The Code of Codes: Scientific and Social Issues in the Human Genome Project.* Cambridge: Harvard University Press, 1992.

Krimsky, Sheldon. *Genetic Alchemy: The Social History of the Recombinant DNA Controversy.* Boston: MIT Press, 1982.

Nossal, G.J.V. and Ross L. Coppel. *Reshaping Life: Key Issues in Genetic Engineering.* Cambridge: Cambridge University Press, 1989.

Suzuki, David and Peter Knudtson. *Genethics: The Ethics of Engineering Life.* Toronto: Stoddart Publishing Co. Ltd., 1988.

Wheale, Peter and Ruth McNally. *Genetic Engineering: Catastrophe or Utopia?* Hemel Hempstead: St. Martin's Press, 1988.

Lee, Thomas F. *Gene Future: The Promise and Perils of the New Biology.* New York: Plenum Press, 1993.

Russo, V.E.A. *Genetic Engineering: Dreams and Nightmares.* New York: W. H. Freeman/Spektrum, 1995.

Internet Resources

You can find plenty of information about biotechnology on the internet by using a search tool such as Yahoo, Lycos, or Alta Vista. Simply enter the words "biotechnology" or "genetic engineering" or a more specific topic that interests you. The following are also good starting points for a wide range of information.

Welcome to Biotech!
http://biotech.chem.indiana.edu/

This interactive site is run by Indiana University as an educational resource and includes a biotechnology dictionary.

About Biotech
http://www.gene.com/ae/AB/index.html

Operated by Access Excellence, this educational site includes articles on the Human Genome Project, issues and ethics, principles of genetic engineering, and state of the art research projects.

WWW virtual library: biotechnology
http://www.webpress.net/interweb/cato/biotech/

An index page directs you to further information about biotechnology, genetic engineering, and pharmaceutical and medical developments.

National Center for Biotechnology Information
http://chemistry.rsc.org/rsc/cba.htm

For the more technically minded, this site includes gene maps and databases as well as a newsletter.

Bio Online
http://www.bio.com/companies/co-info.toc.html

This gets you in touch with a large selection of biotechnology companies and includes details of their latest research projects. Most of these companies welcome online inquiries.

Global Agricultural Biotechnology Association
http://www.lights.com/gaba/online/index.html

A listing of agricultural biotech information available online from various sources.

Biotechnology Information Center
http://www.nal.usda.gov/bic/

Operated by the U.S. Department of Agriculture, this vast site includes directions to patents, press releases, and newsletters, as well as compilations of articles on specific topics such as bovine growth hormone and bioremediation. The site invites questions from users.

Index

Numbers in italic refer to photos or illustrations.

Photo Credits & Sources

Page 10, Don W. Fawcett/Photo Researchers, Inc.

Page 16, A. Barrington Brown/Photo Researchers, Inc.

Page 23, Biophoto Associates/Photo Researchers, Inc.

Page 39, James Holmes, Cellmark Diagnostics/SPL/Photo Researchers, Inc.

Page 42, Prof. Stanley Cohen/SPL/Photo Researchers, Inc.

Page 51, Prof. K. Peter Pauls, Dept. of Crop Science, University of Guelph

Page 111, United States Department of Agriculture (USDA)

Page 113, 140, 183, Harry Turner, National Research Council of Canada (NRC)

Page 117, M.A. Hulme, T.J. Ennis, and A. Lavallée, *Forestry Chronicle*

Page 119, Monsanto

Page 124, Ontario Ministry of Agriculture and Food

Page 144, Boojum Technologies Ltd.

Page 176, Dr. R.H. Devlin, Canadian Dept. Fisheries and Oceans